The Invisible History of
the Human Race

ALSO BY CHRISTINE KENNEALLY

The First Word:
The Search for the Origins of Language

Christine Kenneally

The Invisible History
of the
Human Race

*How DNA and History
Shape Our Identities
and Our Futures*

VIKING

VIKING
Published by the Penguin Group
Penguin Group (USA) LLC
375 Hudson Street
New York, New York 10014

USA | Canada | UK | Ireland | Australia | New Zealand | India | South Africa | China
penguin.com
A Penguin Random House Company

First published by Viking Penguin, a member of Penguin Group (USA) LLC, 2014

LIBRARY OF CONGRESS CATALOGING-IN-PUBLICATION DATA
Kenneally, Christine, author.
The invisible history of the human race: how DNA and history shape our
identities and our futures / Christine Kenneally.
p. cm.
Includes bibliographical references and index.
ISBN 978-0-670-02555-8 (hardback)
I. Title.
[DNLM: 1. Genetics—history. 2. Pedigree. 3. DNA—history.
4. Human Migration—history. QU 11.1]
RB155
616'.042—dc23   2014021679

Printed in the United States of America
1 3 5 7 9 10 8 6 4 2

Set in Bodoni Std Book
Designed by Francesca Belanger

For J&D

You know who you are

CONTENTS

AUTHOR'S NOTE

It was my goal in this book to bring together in a natural way the questions of the general reader about our immediate and historical ancestry and the considerations of the humanities and sciences. The traditional domains have much to offer one another, and this is more true than ever in the age of the personal genome and big DNA data. When I began reporting, I had a memorable conversation with Michael McCormick at Harvard. McCormick is a historian of the Roman Empire and medieval history who has extended his interest in ancient manuscripts to include the Y chromosome, charcoal layers in the soil from ancient fires, and the isotopes in old bones. "What was considered historical evidence in the nineteenth century was only written records; now it's atomic disorder and genes," he said. "The distinction between history and prehistory is blurred, if not dissolved."

I've focused on aspects of inheritance, and our philosophies of it, that either are particularly fresh or have not been paid much attention—their invisibility made them all the more interesting. Herein are studies from psychology, economics, history, and genetics, anecdotes and data from business, science, and the lives of many fascinating individuals. They each exemplify in some way what gets passed down over the generations, and they all provide insights that resonate with one another. As I hope to demonstrate by the end of the book, the concept of ancestry can bring genetics and history together fruitfully; perhaps ancestry will lead us to a place where we can make use of these different kinds of data in a more unified way.

This book focuses primarily on what genetics and genomics can tell us about the connection between an individual's genotype and his history and idiosyncratic traits. Inevitably there are fascinating one-off findings and whole fields of inquiry that I could not include for reasons of length. Twin studies and calculations of heredity based on the differences between them are not referenced here. Before we were able to dig into the genome, twin studies were one of the best ways to try to work out what gets passed down. Identical twins have the same genome, so when they end up looking different, having different aptitudes, or developing different diseases, it tells us that the same genome may be expressed in different ways, and it may tell us what aspects of the environment shape a person via their genome. Twins will always be interesting to geneticists, but traditional twin studies used an indirect method for estimating what is passed down.

Writing a book about DNA presents an inherent challenge, as scientists and scholars of the humanities often think about the subject in different ways while the rest of us may think about it in yet another. Until now there's been little synthesis between the ways we consider genes and health, genes and culture, genes and history, genes and race, and genes and specific traits. We often treat these aspects of our lives as if they were completely distinct, and when we attempt to understand the role that DNA plays, we do so from only one vantage point. But DNA does what it does without regard to any of the conceptual silos in which we attempt to confine it.

It's often the case that when people discuss DNA, they are not talking about DNA at all but rather about the idea of biological determinism, racism, sexism, or the concept of ownership. Or perhaps they're just experiencing a deep-down, angry resistance to the idea that something we do not control may influence our fate. Sometimes people emphasize that the similarities to be found in our DNA prove we are one big family or, alternatively, that the differences in our DNA reveal much about why we are different. Where appropriate I have endeavored to directly address these concerns, and elsewhere I do my best when I use the word "DNA" to mean only DNA.

How DNA shapes us physically is a fascinating but still nascent science. The most solid findings are found in the area of physical health and traits, so I focused on findings from these domains. Does DNA contribute to human behavior, decision making, and complex traits like language and intelligence? Of course it does. But probably not in the ways that many people imagine or fear. At present one of the most complex problems in genetics is how DNA contributes to common diseases. Perhaps once we have solutions to these mysteries, the relationship between genetics and behavior will become clearer.

A note for readers who are scientists, to keep the book accessible, when I talk about locations on the genome, I may talk about "places" or "segments." When I discuss single nucleotide polymorphisms, I often describe them as "letters." Sometimes when I use the word "gene," I am referring to all the possible alleles of that gene. Sometimes I am referring to only one allele, as in "the gene for a particular condition or trait."

Introduction

We follow in the steps of our ancestry, and that cannot be broken.

—Midnight Oil

One cannot and must not try to erase the past merely because it does not fit the present.

—Golda Meir

There's a moment somewhere between the North Pole and the South when I look out my window and see the top of the Himalayas. At first the shapes are clouds, but then they resolve into immense snow-covered crags, perfectly still in the cumulus sea. It doesn't matter how many times I've seen them before; I am always surprised. They are proof that I am somewhere very high and very strange. At thirty thousand feet light floods the plane. Ahead the night drapes across the earth. We barrel toward the darkness, and then we tunnel quickly through it. In a few hours, when we land, it will be morning.

I call my boyfriend's attention to the window. "CB! Look!" He wonders if anyone is over there on the mountains. Perhaps there are a few souls right on the tip. But if so, it's just us and them and all the other long-distance flying containers up here. We are tracing the outer edge of the human envelope. Below, six billion people inhale. Above, there's nothing to breathe.

It's at about this point, when I am too tired to read, I have consumed too much wine, and all I can do is stare, that the human condition starts to get to me. Half the plane is unconscious, the other half is gingerly stepping over them, and there is a permanent line to the toilets. In addition to this aerial trip, my boyfriend and I were embarking on a few heady metaphorical ones as well. We had met in England and were heading to Melbourne to meet my parents. Maybe at some point we would become parents too. It was momentous to contemplate, and it's a truism that we didn't know the half of it, but we didn't even know what we didn't know.

Here's one thing I didn't know when I was flying above the Indian Ocean. One of my great-great-grandfathers was called Michael Deegan, and 170 years before we got on our plane, he followed a similar trajectory, from high in the Northern Hemisphere to way down in the Southern. He was only fifteen years old and he crossed the oceans on a bark called the *Kinnear*. The trip took 105 days. The boat left Dublin with 174 men and arrived in Australia with 172. Deegan had left a country on the edge of a famine in which one million people died and two and a half million people fled within a ten-year period. He had stolen a handkerchief and was now a convict on his way to Van Diemen's Land. Even if he survived his sentence, he was forbidden to return home. All the *Kinnear*'s passengers boarded as criminals, but many of them disembarked compliant. At the end of the journey, the ship's surgeon entered a report on the character of each of the survivors: "good," "orderly," or even "very good." Yet somehow Deegan remained "troublesome."

In the twenty-first century we move through space in these high-speed vessels and through our lives inside little envelopes of time. Usually we know people from the generation or two that come before us, and probably the generation or two that come after, but mortality draws a thick line across the horizon ahead, and existence—or the lack of it—marks the line behind us. It's not common in the West to have met, or even know much about, anyone born three generations before us. Do these people shape us anyway, whether we know about them or not?

Humanity has been making some consequential moves for a long time now. Over the course of hundreds of thousands of years, we populated Africa; then we left the continent in waves and spread across the world. The global travel that began more than sixty thousand years ago continued for the next forty-five thousand years before we settled every habitable continent on earth.

From the ninth century, when Vikings set out to sea in longboats, through the ages of expansion, slavery, the spice trade, and the colonialism that ended in the twentieth century, thousands of ships crossed the oceans. The vessels carried explorers, prisoners, slaves, and immigrants far away from the places where they were born, in most cases never to return. In the second half of the nineteenth century, now known as the age of mass migration, the movement peaked, with more than fifty-five million emigrants embarking from the Old World and heading to the Americas and Australia. On the human scale the ships marked crucial chapters in the lives of many individuals—and in the lives of everyone who descended from them. Culturally they changed the stories of families forever. Biologically the ships delivered mixed samples of human genomes. Everywhere they went they transformed the human landscape, releasing new variants into the local gene pool, creating never-before-seen mixes of human material, and founding novel lineages that branched and then branched again.

I wonder if my convict ancestor thought about whether he would have descendants who might one day look back on his life or who would know him only as a wisp in a cloud of long-dead family. He lived for many, many years after his terrible journey, and though he died long before I was born, I've spoken to people who have spoken to people who once spoke to him. Oh look, here comes the human condition again: He and I have physically touched people who have physically touched each other. But although we will never speak, and I will never see him or hear his voice, he is here with me, and not just in my thoughts. This isn't a metaphor but a fact, as real as the Himalayas. There is information within me that came from him, and if my boyfriend and I have children, some of that information will be inside them too.

We are ourselves vessels. Inside each cell that is inside each person is a massive library of DNA, three billion base pairs that have been passed down to us. I think about this while I sit on the plane; the principle is true for all 466 of my fellow passengers, no matter what class they're in or their reasons for travel. They all carry their great-great-grandparents inside, and they carry traces of their grandparents' ancestors too. Here in this plane are multitudes, 1.4 trillion base pairs that have been passed down through history by millions of people. It's a miracle we ever got off the ground.

In second grade our teacher gave the class a project: We were to go home and interview our parents and then draw a family tree. That afternoon I explained the project to my parents. I would write my name and birth date, and then I would draw a branch growing up out of my name. At the end of that branch I would write their names and what they did for a living, and then I would add their parents' names and birth dates, and I would then draw lines stretching up to them.

I had never before thought much about my forebears. My parents had told me that my father's parents had died before I was born, and my mother's parents lived in another country. And yet here, in the tree I was carefully drawing, I could see that these people had once existed, and they were connected to me.

If I was lucky, my teacher said, my parents would be able to name their grandparents—my great-grandparents—and say when they lived and what they did.

But I was not so lucky. It turned out that my parents were interested in the project, but they were upset about it too: *Why is your teacher asking you these questions? This is homework? What business is it of hers?*

In the end my parents made a few suggestions and, naturally, I wrote them down. They could have kept a lid on their ire and lied to me right then and there, but that wasn't something they did, and they were too indignant to conceal it. So I learned early that the past can bother people in surprising ways and that memory is meaningful but odd. That incident gave rise to a lifelong interest in why family matters, how it shapes you, and especially why people who are long dead still matter so much to you, who are alive.

Pretty much everyone alive today has been asked some form of the question "Who are you?" and the ancient and universal impulse has been to respond by talking about our family. For many years I did a good job of squashing that ancient and universal impulse, studying biology and history in high school and starting a degree and acquiring all sorts of research skills. Occasionally I prodded my mother about my grandparents, but for the most part a thick mist hung in the part of my brain that stores all the questions about family and history and identity. That changed one day in 1990 when my parents and my four siblings were assembled in my parents' kitchen, and my father announced that it was time to tell us that the man we thought was his father wasn't actually his father. The man who raised him, who we thought was our long-dead grandfather, was actually our dad's long-dead grandfather. Our father's mother was our great-grandfather's daughter.

For my father this was an awful admission. He had trouble speaking, and I can still see his crooked right arm and hand supporting his forehead. My mother had her hand on his shoulder, and we five children sat there watching. Their distress was extraordinary, yet none of us shared it, not then and not since. The world has galloped on since my father was born, and at least in our part of it, few people today are concerned with matters of paternity in the way they were in the 1930s. By some mysterious process, even though my conservative parents raised us to be like them, neither did we.

Which is not to say that we weren't disoriented. A truth that none of us had questioned—what is reasonable to call a small foundation stone of our identity—had cracked. Humans tend not to act well when things like that happen, and I still cringe at my first response, which was to exclaim, "I knew there was something going on!" only to realize that my father, who never cried, was choking back tears.

What then? My father didn't want to talk more about it. But I did: What was his name, this man, my actual paternal grandfather, and shouldn't his last name be my last name? Who was he, anyway? Did my father look like him? Wait, did I? Like all normal children, I had spent a considerable amount of time imagining the sudden death of both my parents and my ensuing survival adventure. (My knapsack was packed and ready, just in case.) For half a minute the old daydream stirred, I was long lost to someone else, and here was a truly unknown unknown, a story that hadn't been told.

But my father shut it down again. He had once heard a name, he said, but he had no documentary evidence to prove the connection, so he wouldn't tell us what it was.

And that's where I got stuck. When you put flesh and blood and information together, you are bound to get some heady mysteries and painful feelings, but what can you ever actually *know*?

Our trip past the Himalayas turned out to be the first of many, and CB and I are now married and back in Melbourne, raising two young American Australians that we created from scratch.

This morning, a typical morning, begins in the kitchen in a house not too far from the one in which I grew up. But after breakfast we push the dishes to the other end of the table, and the four of us stare at a container full of salt water, a cold bottle of gin, and a four glasses with a nip of fluorescent green liquid in each. Although it looks as if we are about to make some cheerful cocktails, it is only 10:00 a.m., and our boys are just nine and six. Instead, we are going to extract some cells from our saliva, break them open, and tease the DNA out.

"DNA is—" I begin.

"I know what DNA is," says my nine-year-old.

"Okay. What is it?"

"It's this stuff in your body that goes around." He draws a tight spiral in the air with his finger. "It basically introduces you and where you're from. And it makes your blood type."

I was going to crib Wikipedia, but this pretty much covers it. I didn't learn about DNA until the last year of high school, when we dutifully ran through a few Mendelian tables, adding a dominant gene and a recessive gene or two recessives or two dominants. I enjoyed the neatness of the calculation, but I didn't think much about it beyond that. Back then the word "gene" was barely part of society's vocabulary, but that changed quickly. Starting in the 1990s, scientists and journalists began to announce with ever-growing frequency that the gene for some trait—intelligence, language, red hair, personality—had finally been found. Much has changed since those early days, but many of the ideas that were popularized then are still around in some form today:

- Genes are the atoms of the biological universe.
- Genes are Father Time, predestination, the hand of fate, the story of your life, inescapably inscribed on your cells.

- But only sometimes.
- The connection between a single gene and a single trait may be cleanly observable.
- The connection between a single gene and a single trait may be murky and complicated.
- The connection between many genes and a single trait may be observable and complicated.
- A genetic mutation or a missing gene or an extra gene or a segment of DNA that doesn't seem to make anything may still shape you.
- Or it may do nothing at all.
- Genomes are human bar codes, our ineluctable and most true identity.
- You are so shockingly, so powerfully, so undeniably you that you can't help leaving traces of yourself everywhere you go—a tiny hair, a single cell, a trace of saliva all may lead back to you.

In the late nineties there was even a brief but brightly flaring hope that soon we would all be able to replicate ourselves. Human cloning was a topic in almost every major magazine and newspaper, and at times it felt as if barely a week went by without an article about clones that was written in such urgent tones it seemed our exact replica was about to knock on the front door and demand the car keys. The ethics of cloning were debated endlessly, as were the best candidates for the procedure. In 1998, in a rarely granted interview with the hugely influential software designer and business mogul Bill Gates, Barbara Walters, one of America's most well-known journalists, challenged the billionaire about Microsoft's influence on the world and Gates's personal influence on his company. And then, in a question that must have felt a lot more piercing and relevant at the time, she probed, "Would you want to be cloned?" Wisely, Gates answered that he would not.

As of 2014 there are still no artificially produced human clones, at least none that we know of, and journalists have long since stopped asking people if they want to be cloned. Many well-known animal clones, like Dolly the sheep and the resurrected "extinct" ibex, have died (some prematurely, some horribly). Other cloned animals, like Copycat, the world's first cloned pet (born in 2001), are doing well. (The company that created Copycat, however, is no longer in existence.) In 2013, for the first time ever, a mouse was cloned from only a single drop of blood. In 2014 a Korean company called Sooam Biotech announced the birth of a dachshund clone called Mini Winnie, which it had created for a British woman who won a competition.

If our turn-of-the-century conviction that a race of more "us"es was imminent has proved to be as goofily off-kilter as our 1950s expectation that by now everyone's car would fly, one of the more interesting consequences of this obsession was learning why the idea doesn't actually work very well. We now know that if you took a person's genome and tried to grow another human being from it, the resulting person would, of course, have much in common with the first, but he would be different as well. This is because while you can, at least theoretically, replicate a genome, you cannot replicate the precise conditions in which it developed. Genes respond to the environment around them. They get turned on and off by experience, and any one person's life experience is as unique as our mothers once told us it was. Even identical twins raised in the same household can differ in height, facial features, or any number of traits. Typically the older identical twins get—that is, the more life experience they have—the more different they look too.

As far as science is concerned, it's not just our unique lives that make true clones impossible; it's the myriad unpredictable, unreplicable, and in some ways unknowable elements of a life that may trigger a genetic response. How could we quantify the factors that may affect a single gene in a year? How could we replicate the exact sequence and number of factors that shape one person's genome in a lifetime? What, for example, was the impact on your genome of that unseasonably hot summer when you were eight? What about that traumatic encounter when you were ten? What was the effect of all that bread you ate at school, which was made from that particular wheat, which was imported from another country and which itself had an irreducible, hard-to-recapture life history?

Still, the iconic life of genes—their ubiquitous media presence, the casual way we attribute everything physical or emotional to them or

specifically not to them, the intense attention they receive in the boom-
ing literature of parenthood, their telltale traces at crime scenes, and
their Rosetta stone–like status in the science of being human—has
grown.

CB and I pick up the glasses of salt water and swill it in our mouths. We
are supposed to sluice it for thirty seconds, but the strong taste, combined with the sudden abundance of my spit, makes it hard to hold in.
The saline collects cells from our mouths and as it does, the salt weak-
ens the cell walls. Finally we spit it out into the green liquid, which is
three parts water and one part detergent. "The cells of your body are
like balloons," says CB. "And there's another little balloon inside them,
which is called a nucleus. The detergent bursts the big balloon, and then
it bursts the nucleus too, and then your DNA floats out."

He holds up a glass of saline. "Do you guys want to do it?"

N. is romantic. "Is that what the ocean tastes like?"

F. is unmoved. "No. I'm not doing it. I'm never doing it!"

I offer him two dollars. He swills the saline.

Once the boys have spit out their own saline-plus-cells, CB and I
carefully stir the new mixture. It's still green but foggy, and a white foam
layer has formed on top. Now the salt and detergent are breaking apart
the cells and nuclei. Next we gently add the gin, letting it fall smoothly
onto the back of a teaspoon and into the glass. It creates a new clear
layer between the green and the foam. Because DNA does not dissolve
in alcohol, the idea of this stage is to separate the DNA from the other
bits of cell that are floating in the solution. Then we wait as, slowly, small
white clumps appear, then a spidery line forms on the surface of my liquid. For one soaring second, we are all riveted to the same spot, thinking exactly the same thought: *That's it!*

We sit there poking at the fluid, swirling it around and watching as
the clumps are pulled along, strand by strand. Inside each fragment is
part of a code that has many more bits of information than stars you can
see in the sky. These bits—scientists now say—have the potential to not
just map tens of thousands of years of human genetic history but also to
drop a small pin at the coordinates where our own DNA is located in it.

And here, contained in these four glasses, is the genetic code for my entire family, my beautiful boys, who according to their DNA are not only mine: Half the stuff in their glasses came from me and half from CB, but it also all came *through* us. They are the children of a lineage, and inside the clumps that wander in their soapy water are organized pairs of amino acids that were arranged long before they or I was born—the result of thousands of generations shuffling their DNA.

Can our own personal DNA tell us about the history of the world? Isn't that a bit presumptuous? For most of us, big history is not personal but something out there that we are enjoined to learn, a stack of snapshots of people and things that have been officially declared important.

On top of the pile is everything we think of as modern. Here are an iPhone and a Prius. Here's the conflict in Syria. Here's patient zero with H1N1. Underneath these photos are shots of the glamorous hairstyles, world events, and technological absurdities of the previous decades—here's a cell phone the size of your sneaker, a computer the size of your house.

Further down are images from the nineteenth and eighteenth centuries: industry, ocean voyages, the beginning of journalism. Before these are images of a wonderful time called the Renaissance, when everyone painted and sculpted. Under them there are pictures of the Middle Ages, when peasants wore a lot of burlap and carried a lot of disease.

Prior to the peasants are the Vikings (helmets with nose guards, boats with many oars). They follow grimy snaps of the Dark Ages (muddier burlap, even more terrifying diseases). The pictures of the Romans (roads, plumbing, orgies) come before those of the Vikings, though they follow those of the Greeks (gods, columns, statues). There are a few pictures of biblical times (sand, camels, sun). They sit on top of a long reign of Egyptians (pyramids, mummies, cats).

By the time you get to the mummies, the pile is much shorter. Here are images of a long and undistinguished period of general barbarism, when people wore pelts and carried weapons. Down closer to the bottom of the pile are fuzzy pictures of the Stone Age (stones, scraggly hair, cave drawings), and below these only a handful of pictures remain.

These pictures are so dark you can barely make anything out in them at all. Here's a cave. Here's some ocher. Here, at the bottom of the stack, is a creature that looks a lot like a chimpanzee.

It's fine to carry around these discrete snapshots, but it's worth considering that history is continuous. We can track our way back through it by following the records we create, the writing that goes back six thousand years and the drawing that goes back forty thousand years. We can track our way back with oral histories, rumors, and stories that get passed down. Now, we can also track back through it with the longest thread of all, DNA.

Why do we care about where, or rather whom, we come from? Is it because of what is passed down to us from previous generations? Do the lives of our forebears impact ours in some way? Or is that just a story we like to tell ourselves? Within our families we have many convictions about traits, inheritance, and the mysteries of familial resemblance— she has her father's nose; he has his grandfather's cheeky sense of humor; those two are like peas in a pod. But what is real and what is myth? What do we actually know?

Over the years I've come to view the silence in my own family about my father's father as an actual thing, not as the absence of something. I can see that the secret of his birth had a huge impact on his life, and I suspect it has on mine as well. And what of the man behind the secret? Did he leave anything more significant than the loud bang of a door shut down the generations? About 25 percent of my DNA comes from him. How has it shaped me?

This question led me to a few thousand more questions, which in turn eventually led me toward this book. *The Invisible History of the Human Race* became a personal quest to track down people from all over the world who could shine light on the package of things that gets passed down to us from our ancestors. As it turns out, the package is brimming: stories and secrets, names and dates, feelings, ideas and decision-making tools, and DNA. And that was merely the beginning. What makes the material that gets passed down in both our minds and our bodies so fundamentally interesting is how it shapes our lives, our identities, and our futures.

It was only after I began my quest that I stumbled across the idea of path dependency, a concept invented by physicists to explain how machines could most efficiently use energy to do their work. Path dependency was taken up to fascinating effect by economists to explain how what happens in the future may depend on the path that was taken from the past to the present. They used it to explain why some economies are healthier than others, why some laws are written the way they are, and why technology looks like it does in different places. When we talk about positive or negative spirals in our lives, when something that is good keeps getting better, or when something that is bad feeds back on itself and gets worse, we are talking about patterns that are path dependent.

The QWERTY keyboard is the classic example of path dependency in technology. Of all the keyboard setups that competed to become the standard in the era of the typewriter, QWERTY, for some godforsaken reason, won. Despite the fact that it was not the easiest or the best design for human hands, millions of people ended up using it, and eventually it became so firmly established that it was impossible to revise it. People were committed, and that commitment meant that nearly all laptops and keyboard are today laid out in the QWERTY design.

When I came across the notion of path dependency, I realized like a thunderclap that this book was actually about how path dependency affects people in both their minds and their bodies. Path dependency is, of course, another way of talking about evolution, in that nothing ever evolves completely from out of the blue. The process is fundamentally stepwise: Evolution builds on what came before. The shapes of our bodies and our brains, for example, are constrained by the forms that they took in our ancestors. It's no accident that we look like chimpanzees, because humans and chimpanzees shared an ancestor more than five million years ago.

The question that came to concern me, and that lies at the heart of this book, is how many of our decisions, and how much of our self-knowledge, are ultimately path dependent. The only way to find that out, of course, is to pause, turn around, and look back at all the paths the human race has taken. Ideally that would involve reconstructing some of the significant paths that have led to us to the present—where they began, who traveled along them, where they gently curved, and where they turned more sharply. As the saying goes, it's not really the destination that matters but the journey you take to get there.

Since we are asking what is passed down, we have to ask who is doing the passing. Was it our parents, our society, our government? Has something been passed down incidentally or with purpose? Is the transmission complete or partial? And what do we believe about what we are passing down as we pass it down? Can we even *see* what we are passing down? In the case of culture, we generally believe we can. In the case of DNA, the answer for most of human history has been no.

Still, if you have ever wondered what has been passed down to you, if you have wondered, as I have, how much the past really does matter, you are particularly fortunate to be alive at the present moment. Because all the tools that we need to gaze back at the past—the records we have created, the abstract legacies of culture, and the visibility of DNA—have been radically transformed in the last ten years.

The massive digitization of paper records has completely changed the way we access and use them and, more important, what they can tell us. Additionally, and partly because of the utility of these new digital systems, some canny researchers have worked out how to measure the impact of distant historical events on the attitudes of communities today.

And then, of course, there's DNA, nature's digital record. Much of our general interest in DNA over the past few decades has been in genes and how they affect our health and determine our physical features. But as we got to know the genome better, it turned out that DNA has as much to do with our past as it does with our future. We learned that most of the genome is not coding DNA—that is, genes that express proteins, which then carry out some function in our bodies—but rather noncoding DNA, or what used to be called junk DNA. We now know that even if its impact isn't direct, noncoding DNA may influence our genes in significant ways—or it may do nothing very obvious at all. But even in the latter case, as a group of brilliant scientists has shown, we may learn how to read the book of our history in it.

The most remarkable thing about the use of DNA as a historical tool is that it illuminates not just the biological past but the social past as well. As people make people, who make more people, they pass down their genes and an enormous quantity of noncoding DNA, and in all of this DNA we can trace the choices of populations, as well as fateful personal encounters, that took place thousands of years ago. So, although we have debated for years about the ways in which DNA shapes society (is it deterministic? is it indifferent? does it shape intelligence, behavior, race?), it is far more obviously the case that society shapes DNA.

Now it is becoming increasingly clear that if we bring together DNA with written records and with the more abstract legacies of a community, like its loyalties, emotions, and ideas, what used to be unknowable will come into view in high definition. The continuous streaming of all history becomes more visible, and we can begin to see not just microhistory and macrohistory but also the paths that run between them. What's more, we can begin to untangle the ways in which the operation of genes and our understanding (or misunderstanding) of them also shapes history.

Constructing a history out of all these pieces of information ultimately enables us to understand how the events that occur within our own tiny envelope of existence map onto the stories that extend over long timescales; how lifetimes are shaped by eras and populations; how the lives of ordinary people shape eras; and how much the past makes us who we are and how free are we from it. Using written records, cultural history, and DNA also makes it much easier to understand that the vast forces that shaped the history of the world shaped you and your family as well. And the arrow points both ways: Your family has shaped world history.

But here is the problem with finding out what gets passed down: We often feel strongly that we already know what comes down to us, how it comes down, what it looks like, and how it affects us. First, then, we have to examine our fundamental notions about what gets passed down—which, of course, are themselves passed down.

Part I

Ideas About What Is Passed Down Are Passed Down

Chapter 1

Do Not Ask What Gets Passed Down

Yet why not say what happened?

—Robert Lowell, "Epilogue"

In the lunch line at the local Returned & Services League in Parramatta, Australia, I stood behind two elderly ladies who were comparing notes on their relatives. We were attending a genealogy road show, a modest gathering in a modest town, similar in size and scope to thousands of family history meetings all over the world. At this particular event freelance historians and genealogists-for-hire offered lectures on topics like ship manifests and military history, while archivists advised about document search. The exhibit hall was busy with stalls from companies like Ancestry.com, as well as local societies with an ultrasmall niche, like cricket club records from the 1800s or employee records from the railroad. One group helped convict descendants connect with one another. In Australia being able to link yourself to a convict is something of a genealogical jackpot, much like status in other countries might be conferred by membership in the General Society of Mayflower Descendants, the Descendants of the Illegitimate Sons and Daughters of the Kings of Britain, or Son of a Witch in Massachusetts.

I wandered the display hall, had software explained to me, spoke to experts, and chatted with attendees, but the point of the gathering remained weirdly elusive. A powerful force had brought these people together, but even though they seemed to agree that that force mattered a lot, no one could explain to me precisely how or why.

The ladies in front of me spoke of their lineages as if they were discussing recipes (*You do this, then this, then this*) or perhaps comparing flowers in their gardens (*Oh, you have one of those too!*). The lunch line

was slow, so the comparisons went on and on, and every time they mentioned another person in their respective family trees a bead of sweat
popped up under my hairline. Dear God, here were only two people, yet
they had so many cousins, and they kept piling them on. It wasn't what
they were saying that unnerved me; it was the numerical implications.
Consider: If everyone has four grandparents, then she has eight great-
grandparents, which means she has sixteen great-great-grandparents
and thirty-two great-great-great-grandparents, so that in five generations, everyone has sixty-two people to account for, and that's just the
people in one's direct line.

If you count the siblings of direct ancestors, the numbers are, frankly,
disconcerting. Even if your own parents had no brothers or sisters,
chances are that the further back you go, the more siblings exist in your
family line. It was common in the nineteenth century for families to have
five or eight or even more than ten children, so if you want to explore the
fortunes that have befallen your family in the relatively recent past, you
could easily be looking for three hundred people or more. If you have a
child, his heritage effectively doubles the calculation. In order to trace
the unique tree from which he branched, you must put together your own
crowded lineage with that of your partner's.

This was too much to think about. How could I survey this whole
when it was so paralyzingly big? How could I understand these people
when each of them represented so many thousands of diverse lives? When
my husband was a graduate student, he had to physically brace himself
every time he walked into the university library, as the sight of all those
books, written by all those people, about so many more people, with references to even more people who had written many more books about
other people, was too much to contemplate. Or rather, what was too much
was the multiplicity of the books *plus* the intrusive realization that he
would never in his lifetime be able to read them all. Hundreds of people
blithely stepped over the library's threshold all day long, but for CB, to
catch a glimpse of the library's immense holdings was to come face to
face with his own mortality. I felt much the same at this gathering. How
could so many individuals have existed, each passing something on

before he or she died? All I wanted to do was chat with some keen hobby-ists, but now Death was tapping me on the shoulder in Parramatta.

After a few minutes, though, I started to take in what all the attend-ees were looking at: hundreds and hundreds of historical documents, self-published books, CD-ROMs with lists of lists of names, databases chock-full of people who had once lived and whom no living person now remembered. Everyone in this hall was looking for someone who was gone forever.

Still, even in the presence of all this data, I couldn't seem to find a way to actually get inside the story of family history and world history and the history of all this history, and it wasn't just a problem with this particular road show. Since I had started telling people about the sub-ject of my book, I had received a lot of polite encouragement, but there was a constant buzz of skepticism as well. "Genealogy?" said a friend. "That's like those people who believe in past lives, isn't it? And they're always Cleopatra, never an anonymous slave."

The activity of tracing's one's lineage does not, in fact, have a great repu-tation. One of genealogy's purported offenses is that it has all the real-world verifiability of astrology. That, at least, was a proposition I could explore. But the thing about astrology is that you don't tend to find lots of people being outraged by it. Where genealogy was concerned, I met many people who proclaimed their indifference to it, but it was often an extremely *vigorous* indifference.

"Oh, it's a real American thing," one person observed. "Everyone I spoke to there does it."

Others objected because it was not "real history," being too personal.

Some assumed that because genealogists were not trained archivists, even when they did find a valuable old document with a relative's name or story in it, they didn't know what they were looking at.

Critics with more existential concerns argued that even if you had access to this or that document from someone's past, you could never de-termine from it what he or she was truly like.

In fact, I soon discovered, the criticisms of genealogy were legion.

Genealogists were overly romantic. Genealogists were elitist. Genealogists were divisive. Genealogy, wrote *Guardian* columnist Zoe Williams, "conveys a silent prejudice that never has the guts to announce itself. Ferreting about for antecedents in parish records says, effectively: 'I attach a certain value to having always come from Suffolk or wherever. Oh, no, no, no, I don't mean being foreign is bad, I just mean it's so much nicer not to be.'"

The *Times* journalist Sathnam Sanghera offered some of the most blistering criticisms of family history: "Show me a genealogist and I'll show you someone who is basically obsessed with proving that they come from royal, aristocratic or celebrity lineage. Creepy and boring." Sanghera, who wrote a 336-page book about his own life as a young Sikh growing up in Wolverhampton in the eighties, added: "And before anyone points out the hypocrisy of a memoirist slagging off genealogy, life writing and genealogy are completely different. One being the equivalent of an interest in music, the other the equivalent of an interest in hi-fi equipment. Though perhaps a better way of putting it is that genealogy is the academic equivalent of endlessly googling yourself."

In a blog post titled *Genealogy is Bunk*, science writer Richard Conniff took the argument further. "The rich and famous . . . used to be the only ones who bothered with family history. It was a way to maintain their power, by asserting that they'd always been here and therefore always would be. Curiously, biologists doing long-term studies of primates in the wild say that's how family connections function in nature, too: In high-ranking baboon and vervet monkey families, grandmothers routinely make sure that little Tiffany Baboon and young Percy Vervet III get special treatment from lesser juveniles. This gives the kids a habit of social dominance—and can thus help maintain a monkey dynasty for generations."

Like many critics, Conniff was convinced that in addition to being bunk on the grounds of egalitarianism, genealogy was bunk on evidentiary grounds as well: "Go back ten generations in any family, and the odds are that someone has climbed unacknowledged up the family tree. Sir Winston Churchill prided himself on his descent from the great eighteenth-century general John Churchill, Duke of Marlborough. But the family had a colorful history of sexual misadventure; Winston's own father died of syphilis, and his mother was said to have taken 200 lovers during the course of their marriage." How representative was this privileged woman of the paternity rates in her culture or another? Many critics invoked the specter of illegitimacy as a problem for genealogy.

Conniff explained that his teen daughter was a genealogy buff and that he wanted her to know that "nothing in her genealogy defines her." Really? If a person's genealogy is the series of individuals whose coupling eventually produced that person, then it's hard to see how this assertion is plausible: Surely some part of our identity has come down to us through our parents from our grandparents and from their parents. What we inherited needn't be physical, and it needn't be direct. The way people think about themselves is to some extent a reaction to the *ideas* about identity that were transmitted in their families. If you take genealogy to include all the qualities that characterize the people who are part of your lineage—not just their biologies but also their unique histories (their cultures, their choices, their personalities, and the significant events they lived through)—then the influence that these factors had on them, and in turn had on you, is an open and interesting question. Indeed, the more I considered the case made by the antigenealogists, the less clear it became to me that they knew enough about how these hard-to-define aspects of life are transmitted over the course of generations to conclude that they don't actually shape us.

What about the fact that people are *literally* created from material that comes from their parents and, before that, from their parents' parents? Facial features, for example, are strongly hereditary. What about the fact that how we look often affects how we think about ourselves? I am not arguing that the unique combination of DNA that we inherit from our parents completely, or even mostly, determines who we are or how we see ourselves, but it certainly contributes to it.

While the extent to which we are shaped by the DNA we inherit, and by the degree to which we are influenced by the world in which we grow up (which may largely influence us *through* the DNA we inherit), is contentious, *no one factor* in any part of anyone's life defines him or her. I will return to these ideas later, but for now we can at least say this: If you are someone's biological child, then by definition your DNA came from him or her. You might be raised by people who had nothing to do with your actual conception, and you will no doubt be influenced by their culture. Your genes too will be molded by many aspects of the environment in which you are raised, but your DNA comes to you through your genealogy.

Even if you end up with a version of a gene that neither your mother nor your father possessed, it was created out of the DNA they passed on to you. Even if you end up with a trait that neither of your parents had, your DNA may underpin that trait, and your DNA came from them. Indeed, none of the stuff that is you would hold together in the shape of a human being without the underlying genetic codes, which will continue to function throughout your life.

What was it about family history that inspired so much umbrage? No one rants about self-indulgent knitters or accuses tennis players of being emotionally void. Yet there was a special level of scorn reserved for the world's genealogists. It seemed that a sophisticated worldview precluded thinking about the questions that the Parramatta hobbyists had gathered to ask: What do we have in common with our ancestors? What don't we have in common? What in our lives came to us from them? The more I thought about it, the larger the gap seemed between the real people I met at the road show and the educated consensus on the futility of their quest.

It wasn't just journalists and science writers who dismissed curiosity about one's lineage; academics did too. One professor of psychology told me: "I find it offensive that people fractionate themselves and find meaning in that. It means nothing; it says nothing about who you are." A historian observed, "People only want a family history of themselves that serves their identity at the moment. When it's something ugly, they want it erased."

When I asked an archivist about the attitude in his field, he told me: "It is still a pervasive idea that the genealogists are not our friends, and they're not really users, they're just sort of dilettante hobbyists that are consuming resources."

Even among the geneticists to whom I spoke, those who had looked into their family past were sometimes bashful about admitting it: *It's just a bit of fun, you know?*

In a 2012 essay in the *New York Review of Books*, Richard Lewontin, a Harvard professor of biology, mentioned his role as treasurer for the Marlboro Historical Society in Vermont, where he fields requests for copies of a history of Marlboro written over two hundred years ago by the Reverend Ephraim Newton. The history includes a great deal of genealogical information, and, Lewontin wrote, "Over and over, our correspondents write of the 'pride' they have in descending from these early settlers." He continued:

> Surely pride or shame are appropriate sentiments for actions for which we ourselves are in some way responsible. Why, then, do we feel pride (or shame) for the actions of others over whom we can have had no influence? Do we, in this way, achieve a false modesty or relieve ourselves of the burdens of our own behavior?

Later Lewontin discussed the idea that Jews may share a genetic legacy and that this is considered "a source of group identity and pride." He reiterated:

> Once again we have the question of why having knowledge of remote ancestors and a shared history makes us "proud." Is it that preening ourselves before the glass of history seems less egotistical than inspecting our images in the glass of fashion?

Lewontin's dismissal is unmistakable, and yet by the same argument we probably shouldn't feel any satisfaction when our favorite football team wins a game. Usually we do, though, and perhaps genealogy is no more complicated than that—the impulse to cheer on the home team, even if most of the team is dead. At any rate, if it's not appropriate to have feelings about belonging to a group that is genetically or historically connected, where should the cutoff be? Can we be proud of our grandparents? By Lewontin's measure, no, because we have had no influence over their actions. But the same would go for much of our parents' lives too. There's something nihilistic about asking, *Why should we feel proud?* The implied question is *Why should we feel anything?* Well, why not?

Around the same time that Lewontin's essay was published, obituaries announced the death of Essie Mae Washington-Williams, who was born in 1925 in Aiken, South Carolina. In 2003 Washington-Williams announced at a press conference that she was the illegitimate daughter of Strom Thurmond, a white American congressman, and Carrie Butler, an African American teen who worked as a maid in the Thurmond household. By the time Washington-Williams was in her early twenties, Thurmond was governor of South Carolina. Later he became a long-standing senator in the U.S. Congress and at one time ran for the office of U.S. president. A strict segregationist, he said: "All the laws of Washington and all the bayonets of the Army cannot force the Negro into our homes, our schools, our churches and our places of recreation and amusement."

Washington-Williams waited until Thurmond died before she publicly revealed that he was her father. "My children deserve the right to know from whom, where and what they have come," she said. "I am committed in teaching them and helping them to learn about their past. It is their right to know and understand the rich history of their ancestry, black and white." Washington-Williams was a mother of four, a grandmother of thirteen, and a great-grandmother of four by the time she wrote her memoir in 2005. It's hard to imagine her descendants feeling nothing as they think about her past.

When she finished her historic announcement, the elegant and assured Washington-Williams spoke of the "great sense of peace" that came over her when she decided to share the details of her family's history. For some, no doubt for many, the details of their own lineage, and certainly that of others, may be banal. But when those details are lost or suppressed, they can take on an enormous power.

Perhaps neither question—*Why should we feel proud?* or *Why shouldn't we feel proud?*—can take us far. But it is surely interesting to ask, *Why do we feel proud?* because quite clearly many people do. Even Lewontin allows that all cultures in every era have been interested in ancestry. Were they motivated by pride, sorrow, amusement, or curiosity? It's surprisingly hard to answer this question.

Wendy Roth, a professor of sociology at the University of British Columbia, says that when she was a little girl, she wanted to be a genealogist when she grew up: "I was always the family record keeper." But Roth's family was Jewish, and because of the disruptions of World War II they didn't know much about earlier generations. "There is a sense that other people can do genealogy, but for us the records are gone," Roth explained. "I have one friendship with a woman in England who traced her family back to the 1500s and I've always lived a little vicariously through her."

Once, when Roth was backpacking in Europe, she visited Rymanów, Poland, the town that had been home to one of her great-grandfathers. The only traces of the Jewish community that once lived there were the ruins of a synagogue. Roth met a man who must have been in his eighties, and they communicated in a traveler's mix of drawing and impromptu sign language. She told him she wanted to find the old Jewish cemetery, and he led her up a steep hillside. "It was hard for me to keep up with him," she recalled. "He was so spry." They reached an area of chest-high weeds where the cemetery had been, and he mimed to her that he was a small child during the war. "He acted out that in this cemetery the Nazis had come, and they had lined up Jews, and then opened fire on all of them, and then they took the gravestones and used them to pave a road." Roth's guide then pointed to the road the Nazis had built, which lay beneath them, down the hill.

To feel that she finally stood somewhere that her ancestors had stood was an extraordinary experience for Roth. About fifteen years later, she and her husband traveled to Švenčionys, Lithuania, where, she believed, a great-great-grandfather must have lived. There had once been a vibrant Jewish community there. Unusually, the Jewish cemetery was still intact: "We were crawling around again, it was totally overgrown, lots of stones fallen over but mostly in one piece." They had a translator with them to help find the gravestones for the family, including one for Roth's great-great-grandfather. When Roth saw it, she realized that it also held the name of her great-great-great-grandfather. It was the first time she had learned his name.

It was "unbelievable," she said. "Absolutely unbelievable. I was in tears." Why was this encounter so emotional? "It's a feeling of breaking through a wall. Of the frustration of wanting to know more about your family and your past and what people's lives were like, and where they came from, and who they were, and what their personal stories were, and feeling like you're never going to be able to uncover that. Once you have that feeling of that great mystery, any piece of information feels like a treasure trove. Whether it's a name or a place where somebody once was, it's a form of a connection that you thought you were never going to have."

Roth interviewed many people who traced their family histories (see more on this in chapter 12), and she was always struck by the difficulty they had explaining what exactly compelled them. "With avid genealogists and especially genetic genealogists, I try to get at the question of why they are interested," she said. "What is amazing to me is that people can't articulate it. It's like it is such a basic urge or a basic primal interest that they have a really hard time putting it into words. Some people try, and they'll give you answers that sound a little clichéd. They'll talk about wanting to find their own place in history, or wanting to know where they came from, or they'll talk about it making history more interesting to them, but the level of commitment of a lot of people means to me that it must go deeper than that."

I came across Wendy Roth when I started looking for general studies of genealogy. Someone, I reasoned, must have investigated the psychology of family history or, surely, examined how different philosophies of heredity have shaped this ubiquitous human experience. I asked genealogists, geneticists, historians, and others whose work invoked in some way questions of inheritance and history, but none could point me to a body of work on which he or she relied.

Apart from Roth, a few scholars have investigated the topic, but usually only briefly and in isolation. This seemed odd. After all, historians acknowledge that attitudes and even feelings are passed down culturally. Economists study the way socioeconomic status, especially poverty, is reproduced. Psychologists, social scientists, and even English professors recognize that the family is a powerful engine of inheritance. Most scholars of human behavior accept that individuals are shaped in some way by the people who raised them and that even a family unit may possess a character. Yet there was apparently no field that brought together all these ideas about inheritance.

Like Roth, I became my own family's historian. But even though the fragments of past generations I glimpsed struck me as wondrous, I did not dig down for a long time. When I did begin to look further into my family's past, I experienced a strange array of emotions, including relief and contentment, but there was despair too.

I found myself drawn to one ancestor and then another, and sometimes a single detail was all it took for me to feel, at least for a little while, that I had a relationship with them. The day I realized that I could learn about Julia Dillon, my father's great-grandmother, was a revelation to me. She was one of my sixteen great-grandparents, and I had heard her name many times but knew nothing about her. I couldn't explain why I had never thought about her as an actual person who left physical traces in time.

I found her name on a ship's register and was stunned to learn that in 1862 she brought her four children by herself from Ireland to Australia on a ship called the *Lightning*. At the time I had two small children of my own, and I did everything in my power to avoid taking them with me to the supermarket, where they made a trip through the dairy section feel like *The Odyssey*. How did Julia manage her children on that long voyage?

Once she arrived in Australia, Julia reunited with her husband, Daniel, who had arrived on an earlier voyage. The family lived for a while on the goldfields of rural Victoria, and Julia had five more children. Like so many families of that era, Julia's lost children along the way. Jeremiah died at fifteen of a fever. Johanna died at eighteen while she was in service in a small country town. I found the browned, crisp transcript of the inquest into her death in the state archives. She had been complaining of pain and was sent to bed. Her employer, Elizabeth Farrell, later checked on her. "I found her very hot and racked with pain," she said. "I proposed to give her a mustard plaster." The doctor was summoned and he gave her a dose of medicine and then ordered another dose. "Ought the medicine make her as sick?" wondered Farrell. They left poor Johanna alone, and later Farrell climbed the stairs to look in on her. For a minute she thought she was asleep, but Johanna was gone. "I found her dead," Farrell said. She then called out to the manager that "Johanna was dead in the bed."

Despite the resilience that characterized most of Julia's life, the hardship of her days flattened me. Once, while I was attending a conference in the United States, I woke at 5:00 a.m. with jet lag after covering in a day approximately the same distance that it took Julia forty days to cross. I told a genealogist at the conference that the thought of her hardships made me sad.

"But you can be impressed by her!" the genealogist replied, and I was.

I began to shake the malaise when it occurred to me one day that even though I had all sorts of complicated feelings about Julia, I had no idea what she would have thought of me. Who was I to her? Merely one of her son's many granddaughters. Here is the brutal asymmetry of a big family *and* time. Julia sits at a node in our family tree from which many, many branches sprout. I, on the other hand, am a twig. Even if she were alive, I'm not sure I would mean much to her.

Later I found a photo of Julia. She was about seventy at the time, and it was hard to tell if she was just small in stature or shrunken by age. She wore a frilly black dress and a black bonnet, which was secured under her neck with white flowers arrayed at the front. Her eyelids dragged down at their corners. I suspected the photo had been taken to commemorate a death in the family. Did her eyes, light and inscrutable, reflect sorrow? Was it exhaustion? I still don't know what it means to have a relationship with her, but that was, without a doubt, my father's nose on her face.

How do we personally lose or find information about our lineage? Our tendency is to assume that whatever we don't know about ourselves or our families has simply fallen away naturally, through attrition over time. Certainly, memory has absolute limits, but there are psychological forces at work too.

In 2012 Jordi Quoidbach, Daniel Gilbert, and a colleague wrote about an experiment they had carried out where people in different age groups were asked about either what they had liked, valued, or prioritized ten years earlier or how much they thought their current preferences were likely to change over the next ten years. The scientists found that their subjects were pretty good at assessing how much they had changed in the previous ten years, which was always a considerable amount. By contrast, the participants in the study always underestimated how much they were likely to change in the next ten years. In fact, they didn't think they were going to change much at all.

According to Quoidbach and his colleagues, people have a tendency to think of the present as a "watershed moment at which they have finally become the person they will be for the rest of their lives." The researchers dubbed the phenomenon the "end of history illusion," and they showed that it applied to personality traits, core values, and even best friends. Quoidbach found that the older people got, the less pronounced the illusion was, yet even the oldest subjects believed they had finally become the person they would be for the rest of their lives. "History," wrote the researchers, "is always ending today."

Perhaps the end-of-history illusion also applies to the way people think about generational time. We live in a temporal envelope. For most of us the horizon extends forward maybe two generations and back just two or three. It is hard to break out of the mind-set that we stand at a crucial center point of that span and that all the people who came before were merely precursors to us. It isn't until you populate the family tree out that it becomes clear how brief a human life is, how soon it is over, and how *you* only play a bit part in a story line that expands out and contracts back and goes off in directions that no one can predict or control.

People who dig into their lineage may find it a startling corrective to the feeling that they exist at the apogee of an arc, one that has been heading inexorably toward them and that only gracefully declines away from them. Or maybe people delve into their past because they have started to lose their "presentism" for other reasons. It is telling that the most common triggers for a genealogical quest are major life events, such as the death of a parent or a grandparent or the birth of a child or grandchild, events that typically cause people to stop and consider the more existential questions: *Where do I come from? Where do I go from here? What is my legacy?*

Simply getting older is an incentive as well. "People get married and have children and careers, and then they retire and they have the sad remorse that they don't have their parents around anymore, so they seek for their past," David Lambert from the New England Historic Genealogical Society told me. Aging is, of course, inevitable, and yet in Western culture people are embarrassed by it. In 2012 the novelist Will Self told the *Guardian* that some of the characters in his novel *Umbrella* were inspired by family members in his grandfather's generation. "In a kind of tedious middle-aged way I was doing a bit of family history," he said.

Yet it makes an entirely practical sense that the older you get, the more you begin to perceive the boundaries of your envelope. By the time you reach middle age, you've passed through a few personal epochs, and there is texture in your life history. Chances are the number of dead people in your life has increased too. The consciousness of what is to be lost grows, as does the consciousness of what has been.

While there may be comfort in finding one's place in a big family tree of somewhat similar people, there is disorientation too. I found it dizzying to try to hold all my ancestors in my head, and it wasn't the sheer number of them that made me uncomfortable so much as the realization that all those people once existed, and they undoubtedly all thought that history ended with them too.

My confusion may be explained in part by the psychology of Western culture. A famous study compared the thinking of people from Western, educated, industrialized, rich, and democratic nations (dubbed WEIRD) to people from different cultures. The Westerners, they found, were much more individualistic and perceived themselves to be autonomous and self-contained. They were less motivated to conform and more inclined to feel that they drove their destiny. By contrast, people in non-WEIRD societies were more naturally inclined to see their identity as inextricably connected to their network of family and community. They were enmeshed in roles and relationships and were more oriented to cooperation and the desire to "fit in" rather than "stand out."

Some people may be offended by the idea that they are not entirely in charge of their own destiny, but many—at least by the time they are forty—come to suspect that they are not. If they are older yet and trying to figure out what it means to have a legacy—if their children have children—then simply by virtue of having survived for long enough, they've begun to have a lot more in common with their long-dead ancestors than they used to.

The problem with letting go of our personal presentism is that it not only undermines the sense that *now* is more important than *then* but it also calls into question a whole cluster of connected and comforting assumptions. For example, we experience ourselves as a whole, yet the smallest glance backward tells us we are put together by fractions. While the complicated process of development makes most of the seams of our original self invisible to the naked eye, we were once made up of halves. When our parents made us, they each contributed a sample of their DNA, donating twenty-three chromosomes each; if we take the perspective of our parents' parents, we were made up of four parts, as each of them donated about 25 percent of their DNA to us.

We think of our culture as a whole too. Everything in our day-to-day lives is covered by the same patina of familiarity, but in reality we live in a patchwork of ancient and modern technologies. The same is true of language. Think about a term like "Trojan horse." We apply it to software that smuggles something unwanted onto our computers, but the original concept came from real people who once lived in ancient Greece. What ordinary or new words today might last two thousand years in the future: Dot-com? Internet? YOLO?

Not all cultures regard looking back into one's past as a strange pursuit. Nor do they insist that family history is not also valid history. In fact, the reality that many westerners must actively search for their great-grandparents' names because they don't already know them appears odd to other cultures. Many Asian genealogies are tracked through deep time, and some cultures keep extraordinarily old family record books, adding the name of a new member when he or she is born. These genealogies are descendancy based, as opposed to ours, which are ascendancy based, the difference being that we put ourselves in the spotlight.

"China is where you want to be born if you want to easily do your family history research," one researcher told me. "They have been keeping family history records . . . and have these wonderful family books, *jai pu*, where based on your surname you can go back and trace your ancestry for a couple of thousand years." He continued, "I think westerners sometimes misunderstand the Chinese as *worshipping* their ancestors, but it's really an incredibly deep cultural feeling that if you're not connected to your descendants, your ancestors, and your siblings, you're not . . . it's like you're not whole."

In 2009 an official update charting the descendants of Confucius, "The Top Family on Earth," was completed in China's Shandong Province, where Confucius was born 2,560 years earlier. According to the revised tree, the ancient advocate for peace and social harmony fathered eighty-three generations numbering some two million individual descendants. (An earlier revision in 1937 counted only 600,000 descendants. The new version was the first to include women, overseas family members, and ethnic minorities, like seventy-one-year-old Muslim Kong Xiangxian from Yunnan Province. At the time of the update, the pantheon of Confucius's living offspring included the childless ninety-year-old Kong Demao, the only direct descendent of Confucius in mainland China, and sixteen-year-old James Hung of West London, grandson of a clan elder and a big fan of Manchester United.)

In New Zealand's Maori culture, there is a specialized lexicon for talking about families through time. *Whakapiri* is what Maori do when they work out what ancestors they have in common with others. The recitation of genealogy and stories that are passed from one generation down to the next is called *whakapapa*. It's said that if a Maori learns his genealogy, he will be able to trace his way back through the millennia, person by person, to when his ancestors first arrived in New Zealand.

Many cultures in Africa have an equally strong oral tradition. In West Africa the *griot* is the person who memorizes all the histories and carries the group's identity. In Somalia children under the age of ten learn their genealogy by heart—their lineage, their subfamily, and the larger clan group to which they belong for ten generations back.

Oddly, the critics of genealogy have little to say about the traditional role that the activity plays in other cultures. Yet it's useful to recognize that people all over the world not only think quite deeply about who their ancestors were but also develop different strategies for preserving information about them. One of the biggest assumptions that lurks beneath all the criticism of genealogists is that the decay of information over the years is a natural erosion and that there's something fussy or unnatural about interfering with it. But this is a sweeping assumption to make about a complicated state of affairs. There is nothing inevitable or organic about the informationscape we live in—it is not a landscape.

Quoidbach and Gilbert's work reveals that the way we think about the past is not neutral but involves a psychology of existence and mortality that affects how we see ourselves in time. This psychology must be shaped to some extent by culture, because some cultures embrace their past in ways that westerners typically do not. Clearly, there are huge social forces at work here. Somewhere in the past we have made choices as a society and as individuals to keep some things and to let other things go. Why?

Chapter 2

The History of Family History

Let us now praise famous men, and our fathers that begat us.

—Wisdom of Sirach 44:1

For as long as people have written about genealogy, there have been precocious personal historians who were drawn to the topic at a young age. In the midnineteenth century, Jonathan Brown Bright of Massachusetts complained about his family's lack of interest in its history: "Nobody but myself cares tu [*sic*] pence about it. . . . They are not genealogists constitutionally."

David Allen Lambert was born a century after Bright, and as far I know, they have no family connection, yet they are kin by predisposition. For over twenty years Lambert has worked at the New England Historic Genealogical Society, the oldest genealogical society in the world, but he became interested in his subject when he turned seven. He joined his hometown's local history society when he was eleven, and when at age fourteen he first visited the NEHGS, he was stopped at the door and informed that he had to be accompanied by a parent or guardian. "But they are not interested in genealogy," he explained. He returned when he was seventeen and contributed a thirteen-page report with a hand-drawn portrait of his grandfather on the cover. He also started writing a guide to every cemetery in the Commonwealth of Massachusetts. The book was published years later and is now considered a "cemetery bible," celebrated for its exhaustive inclusion of the smallest gravestones in the most obscure cemeteries of the state. Finally in 2013 he was appointed the chief genealogist at the NEHGS. His original thirteen-page report is still in its archive.

I visited Lambert's street-front alcove in a grand eight-story building on Newbury Street in Boston. He sported a neatly trimmed beard, which was going gray, and, though precise and respectful, he also had a few gentle genealogy jokes up his sleeve. (His grandmother's surname was "Poor" and his grandfather used to say he went over the hill to the Poor house to get a bride.) Among the wooden panels, large chandeliers, and old books we spoke about his own history and the history of genealogy in America.

As a child Lambert found an unfamiliar photograph in the leaves of a book. His grandmother explained that it was of her parents. The idea that his eighty-year-old grandmother had once even had parents, let alone that they had been teenagers during the Civil War, was amazing to Lambert. He went on to discover that his grandmother's uncle, who was blinded in the war, was a drummer boy at Abraham Lincoln's funeral. He discovered that the son of his eighth-grandfather was a judge at the Salem witch trials and indeed was the only magistrate who later recanted. He also found one of the accused in his family tree. According to testimony, his ancestor Lady Mary Bradbury was seen running about her neighbor's yard in the guise of a blue boar. Lambert was eventually able to trace his earliest line to King Cerdic of the West Saxons of Britain, who was born in the fifth century, at least forty-seven generations ago.

When Lambert started his investigating, there was no Internet. "You wrote a letter, sent it to England, and waited a month and a half, and maybe you got a response back, maybe you didn't," he told me. Ever the historian, he also used to go hunting for arrowheads when he was ten. He wanted to know who had dropped them, and he started to research the Indians who lived in his area. One day he went to a powwow, where someone told him, "There goes the chief of the Ponkapoag Indians." Lambert walked up to the man and asked, "Who are you?" The man said his name was Clinton Wixon. Lambert said, "Oh, you're Clarence Wixon's son. Your father was killed by a girl, who had just got her license, while he was riding his bike. That means you're Lydia Tinkham's grandson, and your Tinkhams go into the Bancrofts and to the Burrells. That means you're a Moho through the Momentaug family of the 1600s."

"He looked at me, and his jaw dropped," Lambert recalled. Later the

Ponkapoag made him their tribal historian. When Lambert's parents died, they invited him to a meeting of the tribe where they gave him an Algonquin name that means "one who brought their ancestors back to them that had once been lost, someone who seeks the past." Essentially they call him "Past Finder."

Lambert walked me through the NEHGS building (pausing to ask a guest to move her handbag from John Hancock's chair), and we passed many beautiful nineteenth-century paintings of family groups and extraordinary hand-drawn charts and pedigrees from hundreds of years ago. One 1884 family tree was literally a tree. Etched with fine black lines, three large branches split at the base of a great wooden trunk. Another chart, hundreds of years old and many feet long, was drawn on vellum. Each individual was accompanied by a small, unique coat of arms, which combined and reproduced down the generations: bright golden lions, blue chevrons, and red and white checkerboards mixing again and again. In a state-of-the-art conservation lab, NEHGS staff preserve vellum, antique paper books, and even single sheets of paper, records that people used to carry in their back pockets in colonial times.

Now, as chief genealogist of the NEHGS, Lambert is one of the caretakers of its 2.8 million manuscripts. For him the building is America's attic. "If somebody wrote a letter here in 1897 and sent some photographs or a document, we've had it ever since," he said. Every day, he sits near the attic's front door, and people walk in off the street and say to him, "I want to know about my grandfather. He was in WWII. Where do I find those records?" "My ancestor is a *Mayflower* descendant." "My ancestor was a pilgrim." "My ancestor was on board with Blackbeard."

Genealogy, as we know it, can be traced to the Bible: Abraham begat Isaac who begat Jacob who begat Judas. Around the same time that the Old Testament was written down, Romans painted portraits of their forebears on the walls of atriums, connecting ancestors and descendants with garlands of ribbons. Modern Western genealogy began, of course, with the rise of the aristocracy. The powerful houses of Europe used genealogy to establish lines of succession and fortify dynastic ties through marriage. Many modern genealogies can be tracked back to the 1600s, but only royal lineages—and only few of them—date to as early as the sixth century.

It took many hundreds of years to catch on, but the idea of a lineage became appealing to princely families by the twelfth century, particularly because it was a way of guaranteeing profit from a fiefdom. Members of the petty nobility believed the blood that ran through the veins of their ancestors also ran through theirs, and alongside it in the same direction flowed the wealth. They constructed pedigrees on rolls of parchment that were up to ten meters long. Around four hundred years after that the bourgeoisie adopted the practice.

When Europeans traveled to the New World, they took their ideas about ancestry and their genealogies with them. In colonial America one of genealogy's most important functions was to establish pedigree. Letters from colonists to family in England requested family information, testifying to the common interest in establishing connections. Some colonists sealed their letters with heraldic stamps and had coats of arms incorporated in their portraits or engraved into their silver, while others used English titles or indicated in some other way that they came from somewhere that mattered. Women embroidered family trees. Genealogy at its simplest involved the simple transcription of family names and birth dates in a special book or a Bible. The gravestone of Captain John Fowle, who was buried in 1711 in Charlestown, Massachusetts, bears a coat of arms with a lion, side view, paw raised, and three flowers. A 1658 gravestone in the Abingdon churchyard in Virginia reads:

> *To the lasting memory of Major Lewis Burwell*
> *Of the county of Gloucester in Virginia,*
> *Gentleman, who descended from the*
> *Ancient family of Burwells, of the*
> *Counties of Bedford and Northampton.*

Establishing a connection to power wasn't the only reason for tracing family in colonial America. Benjamin Franklin, born in 1706, was one of the nation's most famous early citizens. An extraordinary polymath, he began his life apprenticed to a printer and went on to found what was at the time America's most read newspaper. He wrote for the paper and other publications (often under a pseudonym) and invented an amazing array of devices, including a musical instrument and an energy-efficient stove, not to mention swim fins. Franklin's kite experiment to explore the nature of lightning has, of course, become a permanent chapter in the history of America, as well as the history of science. He was also an antislavery activist, and in later years he contributed to the American Declaration of Independence, as well as spending time in Paris as a diplomat. He had time for genealogy too.

On a trip to England with his son William, Franklin took a side trip to investigate his family roots in Wellingborough, Acton, and Banbury in Northamptonshire. Throughout his life Franklin identified as a printer, and he remained proud of his working-class origins. Eager simply to inquire about the history of his family, he and William visited cemeteries and read church registers. He established that the first mention of his forebears was the 1563 baptism in Acton Church of Robert Franklin, son of Thomas. Franklin met a number of British cousins, and when he returned to America, he stayed in touch with one of them, Mary Fisher of Wellingborough. If the language of their correspondence is archaic, the tone of the exchange will be familiar to people born in the twentieth century who have contacted distant relatives in another country and shared family stories over a cup of tea. As Franklin wrote to Mary:

> I am the youngest Son of the youngest Son of the youngest Son of the youngest Son for five Generations; . . . had there originally been any Estate in the Family none could have stood a worse Chance for it.

She responded:

> I am the last of my Fathers House remaining in this Country, and . . . cannot hope to continue long in the Land of the Living. . . . I was well pleased to see so fair Hopes of its Continuance in the Younger Branches, in any Part of the World, and on that Account most sincerely wish you and Yours all Health Happiness and Prosperity.

In its most stripped-down form, genealogy was a crucial record of the people to whom you were connected at a time when many families were on the move and society was in a kind of permanent upheaval. Genealogy also memorialized those who passed on. Family genealogists recorded the children who died of scarlet fever or at birth, as well as the successive wives or husbands who from one calamity or another disappeared on the frontier. The 1793 gravestone of Major John Farrar in Shrewsbury, Massachusetts, bears the names of his seven children, "Patty, John, Lucy, Lucy, Patty, Hannah, Releas," nearly all of whom perished within a year of their births; only one child made it to the age of three. In an 1815 letter Leverett Saltonstall told his younger sister that he was creating a record of their ancestors before the information was lost forever, noting, "It is astonishing how little is preserved."

Recording the facts of one's own or an ancestor's life also took on a moral and religious significance for some people. Colonists put down a record of their days in order to inspire future descendants; Puritans believed a "Register of the genealogies of New England's Sons and Daughters" would be used on the last day.

According to François Weil, even from its beginnings American genealogy was a "product of tangled impulses." Weil, who is the chancellor of universities at L'Académie de Paris, is the author of *Family Trees: A History of Genealogy in America*. His curiosity about American genealogy was piqued in 2008 by the intense interest in Barack and Michelle Obama's families when Obama was elected president. The ancestry of America's presidents has always been a topic of general public interest, but the Obamas were the subject of particular attention because they were the first African American first family. Barack Hussein Obama was born in Hawaii to Barack Obama Sr., a black Kenyan government economist, and Stanley Ann Dunham, a white American anthropologist from Kansas. Genealogists tracked Obama's descent from Irish, German, French, Swiss, and mostly English Americans through his mother and from Kenyan Africans through his father. Ongoing research has gleefully noted his family connections to other famous Americans like Sarah Palin, Warren Buffett, Brad Pitt, and even George W. Bush.

When Obama entered the White House, it was thought that because his father was from Kenya he had no links to slavery in the United States. Michelle Obama's family history was thought to be more typical of many Americans because it included black and white Americans, slaves, Confederate soldiers, and preachers. But in 2012 a team from Ancestry.com revealed one of the strangest twists in the story of the Obama family. It turned out that Obama descended from one of the nation's first African slaves, John Punch, via his white mother. A resident of Virginia and Maryland in the mid-1600s, Punch was an indentured servant and, after an escape attempt, was sentenced to a life of servitude.

Clearly, genealogy reveals not just how people build their own identities but also how others view them. Since the election of the forty-fourth president, a fringe political group known as "birthers" has campaigned relentlessly to invalidate Obama's presidency on genealogical grounds. In the face of much evidence to the contrary, birthers claim the president was born in Kenya and therefore has no constitutional right to run for the highest office of the United States.

Despite the amount of attention given to Obama's origins, Weil could find no contemporary account of what genealogy has meant to Americans throughout their history. In fact, he wrote, genealogy is "arguably the element of contemporary American culture about which we know the least." It is striking that the first person to carry out an extensive study of American genealogy is a Frenchman.

Weil set out to catalog America's genealogical motivations over four centuries and he found many. But in certain periods some were more important than others.

One of the most tumultuous periods in American history was, of course, the late eighteenth and early nineteenth centuries, when the growing will to throw off England's yoke and build an independent republic meant that the desire for a citizen to establish noble bona fides became less socially acceptable. As Americans began to reimagine themselves as a nation, the way they imagined their ancestry shifted too. Before and after the Revolutionary War, America's increasingly complicated relationship with England became an increasingly complicated relationship

with time and with the past in particular. This was a turning point—
when genealogy in America became American genealogy. It was also
when modern antigenealogy was born.

Weil documents many signs that the distaste for personal history,
even for mere curiosity about one's family, was born hand in hand with
the new republic. When the Revolutionary War ended in 1783, some officers from the Continental Army formed a fellowship called the Society
of the Cincinnati, named for Lucius Quinctius Cincinnatus, a Roman
statesman. Like many similar groups, the society offered aid and com-
panionship to its members. It also decreed that when a member died, his
membership passed to his eldest son. By the time the society had established a chapter in each of the thirteen states, the membership rules had
provoked uproar throughout the new United States, as the Cincinnati
officers were accused of trying to create a new hereditary aristocracy in
the new republic.

By the time Leverett Saltonstall enlisted his sister's aid in putting
together a genealogy of their family in 1815, he wrote, "These questions
are not for publick information. . . . I should be unwilling it should be
generally known that I have engaged in this inquiry, because it would by
many be attributed to vanity—by all who sprang from obscurity. Vanity
it is not—tho' I confess some pride, and it is a proper feeling."

But what, really, is obscurity? Socially it is lack of status. Literally it
is the absence of a record, and the thing about records is that they tend
to proliferate as a matter of course around people with power: The names
of property owners were recorded in legal documents, whereas early
census takers did not document details about women, slaves, or native
people. The records remain, even as individuals in a family disappear
into the mists of time, and as time goes on, the records take on a power
of their own. If there are no records, there is no power.

Yet people believed that genealogy ran counter to the beautiful idea
that "all men are created equal" and that everyone has a right to "life,
liberty and the pursuit of happiness." Thomas Jefferson, who wrote those
words in the American Declaration of Independence, noted in his 1821
autobiography that his father's side of the family came from Wales and
his mother's side came from England and Scotland. As Weil pointed out,

he also added a humble caveat: "to which let every one ascribe the faith & merit he chooses."

Ralph Waldo Emerson, the essayist and poet, epitomized the aggressively forward-looking character of the new republic when he declared in 1836:

> Our age is retrospective. It builds the sepulchres of the fathers. It writes biographies, histories, and criticism. The foregoing generations beheld God and nature face to face; we, through their eyes. Why should not we also enjoy an original relation to the universe?

Emerson's cry to embrace the present was a call to repudiate the past. His moving meditation continued: "Why should we grope among the dry bones of the past, or put the living generation into masquerade out of its faded wardrobe? The sun shines to-day also." A year earlier he had written somewhat less delicately, "When I talk with a genealogist, I seem to sit up with a corpse."

Weil describes a theater critic who wrote in 1833 that, after watching a play featuring an English baronet who was rather pleased with his own pedigree, the audience left "full of contempt for the English aristocracy, and chuckling at the thought that there are no baronets in America." The author of *Moby Dick*, Herman Melville, whose father, Allan, was rather keen on his genealogical connections to British and Norwegian aristocracy (and whose paternal grandfather notoriously returned from the Boston Tea Party with his shoes full of tea), went out of his way to mock genealogy in his 1852 novel *Pierre; or, The Ambiguities*:

> At the age of fifteen, the ambition of Charles Millthorpe was to be either an orator, or a poet; at any rate, a great genius of one sort or other. He recalled the ancestral Knight, and indignantly spurned the plow.

Critics of genealogy at the time also included foreigners who found America's concern with aristocratic connections odd, if not ridiculous (and many of whom, no doubt, compared Americans' distant claims to lineage with their own more intimate ties). According to Weil, many commented on the particular attachment demonstrated by the upper classes of Philadelphia and Charleston to their aristocratic heritage. One visiting English Tory exclaimed at how "excessively aristocratical and exclusive" Americans were.

David Lambert agreed that for a few particularly influential generations, it was not so acceptable for some people to be too curious about their ancestors. One of the strangest consequences of that era, he believed, is that now he knows more about his grandmother's parents than she ever did. "People back in the nineteenth century were very withholding of information. It was part of that mind-set to look forward instead of looking past."

Certainly the critics were right about one thing—the more people turned to their genealogy to serve a practical purpose in their social lives, especially if it was to elevate their status, the more out of step they became with the spirit of the new nation—and the more vulnerable they became to fraud.

As the United States expanded, the reflex against the idea that the past *must* have meaning for the present developed and spread; at its most extreme it became a belief that the past has no meaning for the present. Nevertheless, it did not quash the compulsion that many people had to look backward. In fact, the genealogical impulse grew vigorously in tandem with the antigenealogy sentiment. What was genealogy *for* if you weren't trying to prove you were aristocratic? In the United States it became an opportunity to prove that you were American. Even as much of the actual practice of genealogy remained unchanged—the keeping of lists in Bibles or commonplace books—it slowly took on a new meaning: In some circles establishing a lineage was no longer considered a badge of superiority, but rather proof of equality. Records were now for everyone, and all families being equal in the great republic, genealogy became an increasingly popular way to honor one's family.

The more time passed after the Revolutionary War, the more people ordered registers and wall hangings that charted their families. The family tree as a symbol became popular, and female students embroidered samplers featuring them. Historical and genealogical societies began to spring up in different states, and some large families even created their own organizations, holding family reunions in New England, Pennsylvania, and New York. One thousand descendants of Robert Cushman, who had organized the leasing of the *Mayflower* in England (and then sailed to Plymouth Rock on the *Fortune* a year after the *Mayflower* arrived) met in Massachusetts in 1855 and pledged to raise a monument to their august history. After an initial period of unpopularity, the Society of the Cincinnati survived to become the oldest hereditary military society in the United States.

Printers produced genealogical registers, magazines were supplemented with blank family trees, publishers printed formal genealogies of particular families, and the first "how-to" books for amateur genealogists were published. Boston even had its own genealogical magazine, and the United States began to produce even more genealogical publications than England. The most important publication in the new American genealogy was John Farmer's *A Genealogical Register of the First Settlers of New England*, first published in 1829, because it became the model for rigorous research. Farmer believed that mere hearsay was not sufficient to prove a family connection, and he advocated a strict adherence to evidence. He corresponded with many antiquarians and genealogists, and the growing community corrected one another's scholarship and began a long conversation about rules of research and proof of lineage. They spoke often of the "science" of genealogy. Indeed, the sternest critics of the bad genealogy of the era were the good genealogists, like John Farmer.

One area where evidence was frequently missing was heraldry, which became more popular in the nineteenth century. For a long time the means to prove genealogical links and claim a coat of arms had resided with heraldic experts or institutions in England, who kept the new genealogy and the new genealogists subordinate to the old. But by the 1850s and 1860s, a heraldic office appeared on Broadway in New York, and American genealogists offered themselves to American families for hire. Not all of them were honest, and even at the time people grumbled about the lack of regulation. Many coats of arms were chosen on a whim from a large catalog of existing patterns or were completely fictitious.

Horatio Gates Somersby, a decorative painter by training, became enthralled by genealogy and heraldry on a trip to England and later became the first London correspondent of the New England Historic Genealogical Society. He traveled throughout the country, transcribing details from formal documents, newspapers, and church records in order to build the genealogical trees of American families. He had many wealthy clients, especially in New England, and, in Weil's account, must surely have been one of the richest genealogists for hire. Yet eventually it became clear that some of Horatio Gates Somersby's research was fabricated.

Indeed, so prolific were his fictions that genealogists today are still being misled. In 1998—more than one hundred years after Somersby's death—the genealogist Paul Reed remarked on a Listserv that Somersby's "frauds have caused me headaches because people descended from lines he forged are not pleased with me for disproving the connections. They want another royal line in place of what does not exist! If the evidence originally existed, he probably would not have had to fake it."

Somersby and his kind didn't just change the pasts of many families; they had an enormous impact on the way the study of history is practiced today. Recall that when François Weil began to research his book he was perplexed to find no chronicles of genealogy in America and he soon uncovered the reason why. It turned out that modern-day historians' aversion to genealogy is part of the foundation of their profession. According to Weil, the 1860s "witnessed the emergence of the first generation of professional academic historians, many of whom took pains to distinguish themselves from genealogists."

In the midnineteenth century there was no huge dividing line between genealogists and antiquarians. But as the cases of counterfeit lineages proliferated and as history became more established in American universities, genealogy was barred from the ivory tower. Partly this was due to its intense popularity. Dixon Ryan Fox, for example, who famously wrote about social history and the economic elite and who taught at Columbia University in the early twentieth century, thought that genealogy developed out of "snobbishness and vanity" and was unworthy of attention.

Truly, the more class-conscious a society is, the more likely it is that genealogy will be used against people of lower classes. For that reason, anyone who cares about equality may view genealogy with suspicion. Yet, while modern society is still affected by social class, what remains of the class system is a fossil of its former self. Dismissing genealogy on the grounds of egalitarianism today is anachronistic, and it ignores the complex emotions of the genealogical impulse. For example, throughout all American history, regardless of how many supporters or detractors genealogy had, Weil notes how often the impulse to record a family's information was provoked by death. In 1829, after the death of his brother, "Daniel Webster, fully conscious that he was 'the sole survivor' of his family, began an autobiography that traced him back to the seventeenth-century colonist Thomas Webster, the 'earliest ancestor' of whom he possessed 'any knowledge.'"

The moral and religious imperatives of genealogy became ever more pronounced with time too. Ancestors were useful for the lessons they provided, good or bad. Eventually the spiritual side of genealogy became an opportunity not just for the living but for the dead as well. In 1805 Joseph Smith was born into a poor farming family in Vermont. Smith claimed that when he was fifteen, two heavenly figures appeared to him and told him that God was unhappy with the world's Christian churches and that he must build the true Church. In 1830 Smith founded the Church of Jesus Christ of Latter-day Saints, also known as the Mormon Church, whose members believe that only membership in the church will save them on the final day of judgment. Lest any members worry about relatives who died before 1830 not having had the opportunity to accept the nineteenth-century teachings of Joseph Smith, the Mormon creed decrees they may retrospectively offer baptism to the departed.

By 1880 Mormon missionaries began to travel throughout the United States to connect with other genealogical groups and transcribe records that could help establish a family connection—and thereby provide a list of possible postmortem converts. In 1894 church members founded the Genealogical Society of Utah, which then planned to build a library devoted solely to genealogical research.

The administration of the government began to demand more record keeping as well. Land warrants and pensions for soldiers or their widows required documentation. In the absence of records, genealogical research was carried out. The first American census took place in 1790 and took account of fewer than four million people. In 1840 the sixth census required 28 clerks to record the demographic details of seventeen million people. By 1860 184 clerks were needed to count the now more than thirty million people in the United States. The census recorded a citizen's name, age, sex, color, birthplace, occupation, marital status, and value of real estate (and whether he or she was deaf, blind, or insane, among other possibilities). With this data the government essentially built a huge set of rudimentary genealogies. The 1862 Homestead Act, which inspired many Americans to apply for a free tract of land out west, generated vast amounts of information as well.

Yet as many of the nation's people began to leave thicker and thicker trails of records, one group had to deal with the fact that all their historical information had been stolen.

Slaves had been brought to America with nothing more than their memories, and after several generations the descendants of those slaves had little chance of confidently tracing the origins of their ancestors. Once in America, however, the lives of African Americans began to be documented, including by deeds of sale; court records; birth, marriage, and other parish records; and military and census data. In contrast to the genealogical curiosity of Americans of northern European descent, it wasn't until late in the twentieth century that African Americans began to study their heritage. A key moment was the publication of *Roots* by Alex Haley in 1976. Allegedly based on a true story, the book was an international success and ignited a passion for personal history in the communities of many minorities, as well as reviving mainstream interest in genealogy.

As it turned out, some of the material in *Roots* has since been disproved, from inaccuracies about the accounts of many characters' lives in Africa (if not their actual existence) to the claim that all African Americans were descended only from slaves. (Just before the Civil War, one in eight black Americans was free.) Still, even if Haley's own claim that he was America's foremost expert on black genealogy has been challenged, *Roots* was a hugely significant book that enlivened history once more for an enormous number of people who had been shut out of it.

Before the twentieth century most documentation concerning heredity was either a personal aid to memory or concerned with legal matters. But in the early twentieth century genealogy became much more connected to biology. It's hard now for us to think of the two as separable, but connecting them was a process of discovery that involved many different minds in many different fields. As scientific ideas were developed about what was passed down, how it was passed down, and what it meant about who we were, they were inevitably shaped by our historical understanding of what is passed down. Thus the initial insights of scientific genealogy reflected the attitudes of the day, like the notion that poverty, talent, and goodness were inherent and—long before anyone began to worry about genetic determinism—the idea that some groups of people were inherently superior to others.

Chapter 3

The Worst Idea in History

We must, if we are to be consistent, and if we're to have a real pedigree herd, mate the best of our men with the best of our women as often as possible, and the inferior men with the inferior women as seldom as possible, and bring up only the offspring of the best.

—Plato, *Republic*

It started with sheep. In the mideighteenth century Robert Bakewell, a gentleman farmer from Dishley Grange, Leicestershire, had a particular talent for noticing what got passed down from a parent to its offspring, and how. For instance, Bakewell realized that specific traits were often linked with families and that the contribution of both ram and ewe had an impact on their offspring. He learned that not only could specific traits be passed down, but sometimes an entire group of traits seemed to be linked, so that the presence of one predicted another. An innocent mark on a sheep's face, for example, might signal that it also had a much more significant trait.

Bakewell began to experimentally breed his sheep and he became skilled at selecting for different traits. Until this point farmers had mostly used an animal's ancestry as a guide to its breeding potential. But Bakewell realized that an individual animal should be evaluated for its own particular set of traits and then methodically bred—or not bred—accordingly. Part of his genius was a knack for amplifying good traits while controlling for bad ones.

At first in secrecy and then to great public acclaim and popularity, Bakewell created a new breed of sheep. Called the Dishley sheep, it had fine bones, fattened up fast, and possessed a strange but wonderful barrel-shaped body whose most valuable parts were larger (while the parts with no market value were smaller).

For generations before Bakewell farmers had bred their sheep for valuable qualities, and with experience they began to develop practical rules for the process. Even though they did not understand that parents pass on specific traits to offspring, they had long recognized that one animal could be of more value than another and that the best animals had great currency in trade and war. The basic rule of thumb, which had been around since the time of the Greeks, was "like begets like" (still a reasonable guide today). But even though their experiences had enabled farmers to formulate reliable axioms for producing good stock, they did not create any new breeds. At the time, they didn't conceive of the process of selection and reproduction of stock over generations as "heredity"; rather, they envisioned it as a holistic process and spoke about the way sires might "leave an impression" or make a "stamp" on offspring.

The understanding of reproduction at the time was still heavily influenced by the Bible's version: Humans are "fashioned to be flesh . . . , being compacted in blood, of the seed of man." Some farmers believed that traits were passed down by blood, and some thought that particles in the blood collected in the testicles and were somehow turned into seed. The basic idea was that beings were not *reproduced*; they were *created*. Therefore an individual animal was shaped by its ancestry but also by the weather, its food, or even its dreams. The most delicate moment in the creation of an animal was the moment of its conception. Even what the mother was looking at when the animal was conceived could shape it.

For a long time farmers believed that the health of stock was so closely connected to its environment that if you moved it from place to place, it would degenerate and its value would suffer. It was thought as well that males and females made different contributions to the creation of a new being. Some theorists attributed most of the creative power to the egg (which was woken up by the sperm) or, more typically, to the sperm, which planted the stuff of life in the egg. Bakewell's experiments made it clear that the female was as important as the male when it came to breeding.

In 1783 Bakewell founded an association to regulate the leasing of the Dishley to other farmers for breeding. It was the first time that a farmer systematically leased stock based on its breed (and charged revolutionary prices for the privilege). Bakewell's experiments soon gave rise to a large collective activity, and the combined genius—and stock— of his neighbors and, later, of most of sheep-breeding society changed what we know about the way that traits move through generations.

Bakewell became known as the "Prince of Breeders," and his Dishley sheep made their way throughout England and on to Europe and America, eventually being bred in Australia and New Zealand as well. By 1790 one of his contemporaries made the modern-sounding observation that Bakewell's experiments showed that "a number of traits were found, in some considerable degree at least, to be hereditary." That principle didn't apply only to sheep, as Bakewell also bred cattle and horses that became extremely popular with other breeders. Before his experiments farmers spoke of characteristics that remained "constant" and true" over generations; afterward "inheritance" became accepted as a fundamental mechanism. In 1915, more than a hundred years after his death, the *Breeder's Gazette*, the most widely read breeding publication in the world, wrote:

> Flying squarely in the face of all preconceived notions governing the production of farm animals, he was the first of the world's great animals breeders, [demonstrating] the readiest and most effective method of establishing and fixing desired characteristics.

Bakewell's keen powers of observation and his systematic approach forever changed the way people thought about what gets passed down— and the extent to which it can be controlled. But even though he developed masterful techniques for manipulating heredity, he didn't understand its mechanics. That would take another century.

After sheep breeders, the second-most-important group in the history of heredity was the French medical community. Before the early nineteenth century *hérédité* was primarily a legal term used with reference to inheritance and bloodlines, but around 1830 French doctors began to think about the hereditary transmission of physical features through families and to use the word in a biological context. After 1840 doctors also considered the possibility of moral or psychological traits being passed down.

By the end of the nineteenth century, doctors—and increasingly other scientists in the life sciences—came to agree that heredity might explain a whole set of phenomena that had previously been thought to be entirely unrelated, like the recurrence of disease and resemblance in families, the differences among races, and even the formation of species. It was possible for the first time to talk about traits and the connections between them without also speaking of the specific individuals who possessed those traits.

At around the same time, Gregor Mendel became the first person to figure out how this process actually worked, or at least how part of it worked. Born in 1822 in northern Moravia (now in the Czech Republic), Mendel grew up toiling in his family's orchards. In 1843 he entered the monastery of St. Thomas in Brno and began to work for Abbot Cyril Napp, who was head of the Moravian Agricultural Society, as well as a member of a number of other agricultural and scientific societies. Napp sent Mendel to study at the University of Vienna for two years. On his return Mendel was asked to look after the monastery's garden, where he began a series of studies of pea plants, investigating the transmission of characteristics between generations, applying pollen to the plants himself with a small paintbrush.

Mendel experimented with height, color, seed texture, and other features, and he concluded that for certain traits offspring received something from each parent that contributed to the trait. The elements that were passed down were either dominant or recessive, meaning that if one parent passed on a dominant version of a trait (like smooth skin) and the other parent passed on a recessive version (like wrinkled skin), then the dominant would always appear in the offspring. If both parents passed on the dominant trait, then that would also appear in the offspring. Only if both parents passed on the recessive trait would that characteristic emerge in the offspring. If parents with a dominant and recessive element had four offspring, the probability was that three would have the dominant trait and only one would have the recessive.

Mendel's theory explained how children may possess a trait that is the same as one parent's but not the other, and why some traits seem to skip a generation. If an individual received the dominant element from one parent and the recessive element from the other, he will express the dominant element, but he might pass the recessive version on to his own offspring. If that offspring also received a recessive element from its other parent, then it might look more like one of its grandparents than either parent.

Mendel published a paper about his findings in 1866, but it had little impact; he, much like Bakewell, was so ahead of the preconceived notions of the day that the significance of his insights was not appreciated. It wasn't until three decades later that scientists came to realize that Mendel had neatly outlined some of the key principles of heredity. In 1906 the English scientist William Bateson first used the term "genetics" to describe work based on those principles.

For all the individual contributions, collaboration, experimentation, and inspiration that it took to develop the idea of heredity, it was only the first idea of three that would radically change the way people thought about generations and genealogy in the late nineteenth century. The second idea was, of course, evolution.

It's hard to overstate the impact that the concept of evolution has had on all science, medicine, conservation, and social sciences and on most of modern-day life. When Darwin's *On the Origin of Species* was published in 1859, its definitive dismissal of a divine creation caused a furor. Darwin proposed that the individuals within a species naturally varied from one another. They adapted to their environment, and the animals that were best adapted reproduced in greater numbers. Darwin's theory made it possible for science to think beyond the genealogy of the human race and imagine an unbroken line of mothers, beginning at one end with a creature that looked like a chimpanzee and continuing down through the line—each mother giving birth to and caring deeply for her child—to stop at the other end with us. Still, while Darwin knew that something *like* genes must be at work in the creation of individuals and species, he didn't know what that actual material was.

Darwin was influenced by the French discussion of *hérédité*, and he later even offered his own theory of hereditary transmission, which he called pangenesis. He proposed that small particles, called gemmules, were passed from parent to offspring and that they accumulated and helped shape traits. Although it was a flawed theory and was never seriously taken up, Darwin understood that heredity needed to explain how traits that weren't apparent in a parent might appear in a child. Sadly, he never became aware of Mendel and his experiments.

When Mendel's discovery was finally taken up, it was with great and universal enthusiasm, because it so eloquently explained an observable phenomenon and helped breeders make reliable predictions. It also appealed because of its neat calculus, one that made something that had once been so mysterious now seem so controllable. The intoxicating feeling of mastery, plus the grand vision that evolution provided, seemed to lead to the tantalizing conclusion that men should take charge of the process. If artificial selection could be used to create better cattle, horses, and sheep, then why not create better humans as well? Eugenics was the third idea to completely transform how we conceive of generations and what is passed between them.

Francis Galton, Charles Darwin's first cousin, thought that breeding better humans was an excellent idea, and thus the science of eugenics, a term that Galton coined, was born as a twin to the science of heredity. Galton enthusiastically conducted wide-ranging inquiries into heredity and many other fields. Like many other scientists at the time he was able to fund his own experiments through his family's wealth, and he ranged far and wide geographically as well as intellectually. Galton became a rather famous explorer of South Africa and an influential early student of meteorology, and he worked out a method of categorization of fingerprints that is still used today. He was also the first person to study heredity in twins and develop a formula for predicting how much of a trait is inherited.

Galton was profoundly influenced by Darwin's ideas, as well as by his success. To test the notion that gemmules circulated in blood, Galton transfused blood between unrelated rabbits to see if traits could be passed on that way, an experiment that proved unsuccessful.

Still, Galton was enthusiastic about the idea that artificial selection— that is to say, breeding—could weed out bad qualities and foster good ones in humans. Indeed, he believed that many traits were heritable in a straightforward way. If your father was brilliant, you had a good chance of being brilliant. If your father had a weak constitution and was fragile, then you probably would be too. If your parents were unsuccessful, then you would almost certainly be as well. Galton—who coined the phrase "nature and nurture"—also believed that biological traits were fixed, determinants of an inescapable fate that could only marginally be influenced by education or other social forces. To prove his theory, Galton assembled a list of twenty-five hundred eminent men and tracked family relationships on that list. He found that men on the list were related to each far more often than could have occurred by chance. Galton's interpretation was that genius underlay this eminence, and that his list proved genius was heritable

The basic spirit of eugenics, then, involved grafting the ideas of breeds, traits, and improvement onto preexisting social divisions. In Galton's Victorian world those divisions were thought to be natural, and it therefore made sense to him to look for biological rather than social explanations for the inequity in people's lives. When conditions like poverty, criminality, and insanity were considered to be "natural," it made sense to try to deal with them on the biological front. Why not breed them out? "What nature does blindly, slowly, and ruthlessly, man may do providently, quickly, and kindly," wrote Galton, who believed that eugenics could be a new religion.

Galton's ideas spread across the world, where they were quickly taken up in the newly energized post–Civil War America. There the egalitarian spirit was utopian but selective: Everyone was equal in the new democracy, except for women, black people, people with disabilities, and the poor.

Madison Grant was born in 1865 to a family that was well aware of its ancestry. On his mother's side Grant descended from a Walloon Huguenot who had settled the "New Netherlands" in 1623. Grant's father was a well-known Newark physician who had many fascinating ancestors, including the Puritan Richard Treat, who settled New England in 1630.

Raised in Murray Hill, Manhattan, Grant was educated by tutors and spent his summers with his three siblings at their grandfather's Long

Island estate. As a teenager he spent four years in Europe receiving a private education and visiting museums. As an adult he became a member of the Society of Colonial Wars, whose exclusive membership was open only to male descendants of high-level participants in the conflicts that took place between 1607 and 1763. Members of the society received a certificate that detailed their family histories, and each year the society published a yearbook that included the genealogies of all its members. Grant was extremely bright, worked hard, and became one of America's first and most powerful conservationists, as well as its "most influential racist."

The first biography of Grant, published only in 2009, was a project made difficult by the fact that after he died in 1937, his family destroyed his papers. Grant's biographer, Jonathan Peter Spiro, spent years combing archives for traces of his subject. He noted that even Grant's wide circle of friends seemed to have disposed of any documents that mentioned him. Nevertheless, Spiro identified a few key moments in Grant's intellectual development.

On his educational tour of Europe Grant visited the Moritzburg Castle, a Baroque hunting lodge. There, Spiro imagines, he was transfixed by the lodge's extraordinary collection of red deer antlers, which had come from animals hunted three hundred years earlier. In a grand dining hall that can still be visited, the walls are covered with antlers extending two stories high, and at six feet six inches across, one of the pairs on display remains the largest set in the world today. It would have been apparent to Grant that these antlers were all much larger than those sported by most of the deer he had hunted. The red deer, Grant would have concluded, was degenerating.

Another of the privileged societies to which Grant belonged was the Boone and Crockett Club, a group of self-styled adventurers "who believed that the hardier and manlier the sport is, the more attractive it is." Grant was close to the club's founder, Theodore Roosevelt, who went on to become the twenty-sixth president of the United States. Grant and the gentlemen of the club spent a great deal of time hunting. Their extensive, intimate contact with wild animals led them to realize that the animals of North America were getting smaller and shrinking in number.

Inspired by his love of nature and his fear that it was being changed forever, Grant exercised his considerable social power to extraordinary ends. He lobbied to save the American bison from extinction. He cofounded a Save the Redwoods League to ensure that the giant California trees were not all chopped down. He was passionate about conserving whales and bald eagles, among other animals, and he was one of the founders of Glacier and Mount McKinley (now Denali) national parks. Grant was also a founder of the New York Zoological Society and the Bronx Zoo. At the turn of the century he was instrumental in the creation of a number of exhibits at the zoo, including one that featured a man from Africa.

Ota Benga was a Mbuti pygmy from the Congo. He was four feet eight inches tall, and his teeth had been filed to points. Even before he crossed paths with Grant, Benga had experienced much tragedy. He was married with two children, but one day when he was hunting, his family was murdered by King Leopold's Force Publique. Later he was captured by slave traders and purchased by a missionary for a bolt of cloth and a pound of salt. The missionary had come to Africa specifically to acquire a selection of pygmies to display at the St. Louis World's Fair. After his stint as an exhibit, Ota Benga returned to Africa, but he felt that he no longer belonged there and he went back to America. For a while he was on display at the American Museum of Natural History but soon ended up spending time in the monkey cage at the Bronx Zoo. A sign outside Benga's display read:

The African Pygmy, "Ota Benga."
Age, 23 years. Height, 4 feet 11 inches.
Weight, 103 pounds. Brought from the
Kasai River, Congo Free State. South Central Africa,
By Dr. Samuel P. Verner.
Exhibited each afternoon during September

Soon after he became a popular public spectacle, a delegation of "colored ministers" from the Colored Orphan Asylum in Brooklyn approached Madison Grant to plead the pygmy's case. Grant, who was apparently quite charming, reassured them that Benga would soon be leaving and that while he was at the zoo he was helping look after its animals. That afternoon the delegation, accompanied by reporters, returned to the zoo and found Benga locked with a guinea pig in a cage, outside of which hundreds of people stood gawking.

Benga was allowed to walk the grounds but always under the watchful eye of a groundsman and even police. Eventually he was released into the care of the Colored Orphan Asylum and later sent to Virginia, where he made plans once again to return to Africa. But he never went, and after years of working in a tobacco factory, Benga committed suicide by shooting himself in the heart.

In today's world, where conservation is considered a necessity and a virtue and racism is regarded as deplorable, Grant is a hard person to understand. But for him, preserving his beloved redwoods and bison, putting human beings on display, and saving the Nordic race were all part of the same package. Grant believed that all these actions were a benevolent form of stewardship.

Historians trace the xenophobia of Grant's era to the 1880s, when U.S. immigration jumped from 250,000 new people a year to over half a million. Earlier immigrants had been predominately from northern European countries, like Germany, Britain, and Ireland, and it's true that some of them were less welcome than others. In the 1850s, prejudice against the Irish fueled an anti-immigration movement. But a few decades later, more and more immigrants began to come from the other parts of the continent. The latter groups lacked urban skills and formal education, and they began to fill up the cities of the eastern United States. Contemporaries of Grant wrote about how anarchic it felt to be on the streets of New York, which were filled with crowds of European peasants. Unemployment was rife, and crime and poverty were out of control. The newcomers could not have been more different from Grant and his fellow northern European Americans and seemed to threaten every aspect of Grant's privileged world. He wrote, "The immigrant laborers are now breeding out their masters and killing by filth and by crowding as effectively as by the sword."

Grant rued the fact that Americans had brought their destruction upon themselves: "It was the upper classes who encouraged the introduction of immigrant labor to work American factories and mines. . . . The farming and artisan classes of America did not take alarm until it was too late and they are now seriously threatened with extermination in many parts of the country." He likened the situation to the fall of Rome, where the lower classes succumbed first, but the "patricians" were taken down a few generations later.

Still, Grant was born a few years after Charles Darwin's *On the Origin of Species* was published, and while his parents' generation was deeply shocked by the idea that nature—not God—selected who would live and who would not, Grant was invigorated by the notion that humanity could direct selection by controlling traits passed down in families. Any sheep breeder knows, Grant wrote, that apart from the occasional throwback, black sheep have been bred out of domestic herds by selectively not breeding them.

If all "social failures" were sterilized, Grant argued, humanity could rid itself of the unfit. "This is a practical, merciful and inevitable solution of the whole problem," he wrote, "and can be applied to an ever widening circle of social discards, beginning always with the criminal, the diseased and the insane and extending gradually to types which may be called weaklings rather than defectives and perhaps ultimately to worthless race types." While Grant agreed that the state should nurture the defective individual, it was the state's duty to ensure that his "line stops with him."

Even by the standards of his day, the theory of eugenics was shot through with basic illogic. Grant wrote:

> The Negroes of the United States while stationary, were not a serious drag on civilization until in the last century they were given the rights of citizenship and were incorporated in the body politic. These Negroes brought with them no language or religion or customs of their own which persisted but adopted all these elements of environment from the dominant race, taking the names of their Masters.

After more than one hundred years of abduction, abuse, and slavery, the absence of a native culture (and, implicitly, records) in the African

American population represented for Grant a sign of African inferiority. The key point for him was that the African American adoption of Anglo-Saxon culture proved its superiority.

Similarly, Grant's eugenics was built on a muddled version of evolution and heredity. These new ideas were being imperfectly applied in a world that had for all of recorded history been committed to the idea that some humans were superior to others. Darwin's theory, Mendel's science, and the implicit use of genealogy were never the real problem; it was the way they were used to give long-standing social divisions a scientific rationale. The world in which these men lived was already one of great inequity, where poor people were considered to be poor because of their own inadequacies, not because of society's. Even Darwin made a distinction between "savage" and "civilized" races, and this was typical for his time.

At the core of Grant's fears and his ideology was the idea of purity and the way in which it could be tainted. This was a common preoccupation of the day, and it was compounded by the 1918 flu epidemic, which killed twenty-one million people all over the world, creating much anxiety about contagion and hygiene.

The notion of racial purity was enshrined by law in many American states, particularly in the South, where legislation was used to segregate black people on public transportation and in schools, public restrooms, and other public places. In its most extreme version, known popularly as the "one drop" rule, race was reduced to a formula based on parentage, as Grant explained: "The cross between a white man and an Indian is an Indian; the cross between a white man and a Negro is a Negro; the cross between a white man and a Hindu is a Hindu; and the cross between any of the three European races and a Jew is a Jew." Grant invented the term "the Nordic race" to describe his own blue-eyed, light-haired stock.

In 1916 Grant outlined his philosophies in *The Passing of the Great Race*. The book was translated into German in 1925 and was much quoted by scientists in the eugenics movement there. It was also read by a young Adolf Hitler, who—so goes the story—later wrote a fan letter to Grant to tell him that *The Passing of the Great Race* was his Bible.

Madison Grant remains the darkest and most disturbing figure of turn-of-the-century American genealogy and eugenics, but the movement didn't end with him. At the beginning of the twentieth century, eugenics societies in many states were formed to try to influence government and promote the betterment of humanity. The Eugenics Society of America held Fitter Family competitions at state fairs, which families entered by undergoing psychometric, dental, and other examinations and by filling in a Fitter Families examination form listing their "physical, mental or temperamental defects" and their "special talents, gifts, tastes or superior qualities."

The competitions were a little less stringent than such qualifications might suggest. In Massachusetts in 1925 a family that won the "average family cup" admitted to suffering from myopia but fortunately numbered among their special gifts mathematics, languages, literacy, and golf. The winners received a medal on which the classical, athletic-looking figures of a man and woman in flowing robes stretched out their arms to a naked toddler; above them was the inscription "Yea, I have a goodly heritage."

Of course, the celebratory focus on breeding good traits went hand in hand with concerns about not only the flood of bad traits that accompanied the huge influx of immigrants but also the declining rate of childbirth in the middle and upper classes ("the most valuable classes," according to Madison Grant). One eugenics poster at a state fair asked: "How long are we Americans to be so careful for the pedigree of our pigs and chickens and cattle and then leave the <u>ancestry</u> <u>of</u> <u>our</u> <u>children</u> to chance, or to 'blind' sentiment?"

Certainly the tendency to confuse social conditions with "essential traits" and then give them a biological explanation looked even more scientific when it was placed in a Mendelian framework. A poster about "fit" and "unfit" marriages presented the equation like this:

Pure + Pure: Children Normal,
Abnormal + Abnormal: Children Abnormal
Pure + Abnormal: Children Normal But Tainted,
Some Grandchildren Abnormal

**Tainted + Abnormal: Children 1/2 Normal But Tainted,
1/2 Abnormal,
Tainted + Pure: Children 1/2 Pure Normal, 1/2 Normal
But Tainted
Tainted + Tainted Children: Of Every Four, 1 Abnormal,
1 Pure Normal, And 2 Tainted**

So closely knit were the ideas of trait selection in humans and in animals that the many subcommittees of the American Breeders Association, whose interests spanned humans and animals, included the Committee for Heredity of Insanity, the Committee for Heredity of Eye Defects, the Committee for Heredity of Criminality, and the Committee for Immigration. The association published a magazine, which in the third issue of 1912 included articles like "Domestication of the Fox," "The Breeding of Winter Barleys," and "Heredity of Feeblemindedness." In a piece titled "A Study in Eugenic Genealogy," the writer spoke of vigor and virtue as dominant Mendelian traits and weakness and vice as recessive. The journal included detailed family trees to illustrate the transmission of "defective strains," which included traits like epilepsy, insanity, juvenile delinquency, and wanderlust/vagabondism.

The neatness of Mendelian explanations—the way true pitch seemed innate, the way color blindness was genetic—made people giddy with its possibilities. Alexander Graham Bell, best known for inventing the telephone, was passionately interested in the application of heredity and eugenics but worried that the public was put off by the eugenics movement's emphasis on the negative and advised the pursuit of positive traits. He opened a genealogical records office in his laboratory and studied hereditary deafness in Martha's Vineyard. He became chairman of the Eugenics Records Office in New York, which was devoted to studying human longevity. Bell and his workers carefully examined the lineages of families with members who lived to be over eighty.

Paul Popenoe, a close associate of Madison Grant, wrote extensively about America's great eugenic future, in which genealogy would play a crucial role as nothing less than the "handmaid of evolution." The traits recorded in a pedigree were not just "personal matters," he argued;

"upon such traits society is built; good or bad they determine the fate of our society."

In addition to birth and death dates, names, and relationships, Popenoe advised that genealogies should also include notations of traits, talents, and flaws. Americans should send their genealogies to a central office, where researchers could use the material to learn more about heredity and apply their findings to medicine, law, sociology, and statistics. "The alliance between eugenics and genealogy is so logical that it can not be put off much longer," he wrote.

Genealogists understood themselves better, experienced broader outlooks, and led worthier lives, claimed Popenoe. Society could use genealogies to make decisions about the education of children, who could be directed based on their inborn abilities (which we now know meant not the child's abilities at all but the abilities that had been previously measured in the parents).

Remarkably, Popenoe moved on from eugenics to invent the field of marriage counseling, in which he later worked exclusively, advising people to first choose marriage candidates based on genealogical information and only then fall in love. (A 1925 book reviewer summed up the approach: "Keep a card index of all the candidates . . . draw up a list of wifely qualifications . . . Health, Motherhood, Intelligence, Appearance, Homemaking, Disposition, Age, Family, Vivacity, Comradeship . . . assign a value of ten points to each of these . . ." Eliminate anyone whose total is less than seventy-five points.)

From a distance Popenoe's program is unfeeling, mechanical, and, for all its seeming pragmatism, completely impractical. But there was grandeur in his vision too, and the way that he thought about genealogy and relationships prefigured the social-network thinking of the early twenty-first century. Genealogists saw their families not just as an "exclusive entity, centered in a name dependent on some illustrious man or men of the past; but rather as an integral part of the great fabric of human life, its warp and woof continuous from the dawn of creation and criss-crossed at each generation." Genealogy helped people gain perspective on "the sacred thread of immortality, of which they were for a time custodian."

Unfortunately, some threads were more sacred than others, and if the great warp and woof of the human fabric was tattered at its edges, Popenoe advised trimming those edges off. He examined the question of the extermination of dysgenic people ("From an historical point of view, the first method which presents itself is execution. . . . Its value in keeping up the standard of the race should not be underestimated.") although in the end ruled it out: "To put to death defectives or delinquents is wholly out of accord with the spirit of the times, and is not seriously considered by the eugenics movement."

Popenoe, like Grant, did advocate segregation and, as a useful next step for special cases, sterilization. That remedy was taken up by many states, beginning in Indiana in 1907. The first American to be forcibly surgically sterilized was seventeen-year-old Carrie Buck, who was described as belonging to the "shiftless, ignorant, and worthless class of anti-social whites in the South." Buck had one child, who had been conceived as a result of rape. Despite this, and the fact that her first-grade school report revealed a B average, she was described by the men who decided to sterilize her as promiscuous and feebleminded. Others who were subjected to this process included criminals, the blind and the deaf, orphans, the very poor, and people who were institutionalized. Between 1907 and the 1970s at least sixty thousand Americans were judged inadequate and forcibly sterilized by their states. The consequences are still playing out. In 2014, residents of North Carolina who had been sterilized by the state between 1929 and 1974 were able to file for reparation from a state fund.

Eugenics was taken up with great enthusiasm in Norway, Austria, Denmark, Finland, Belgium, Russia, France, Mexico, Brazil, and Japan. Many nations sought to bureaucratize human breeding and control heredity by establishing federal offices to track genealogy, like the Eugenics Record Office (ERO) in the United States and the Swedish State Institute for Race Biological Investigation.

Sterilization policies were implemented throughout the world, and everywhere the spread of eugenics was considered part of the grand project of bringing science to the masses. Much as they had been featured at American state fairs, science and eugenics were presented in other countries as a combination of education and entertainment. Women competed in eugenic beauty contests and attended hygiene exhibitions.

In Japan in the 1920s, eugenic marriage-counseling centers were opened in many cities, some in department stores so that shoppers could browse clothing, home goods, *and* the latest news about race hygiene. Eugenic marriage counselors served as matchmakers too, bringing couples together if they had the right genealogy and health certification. In the absence of a certifiable match, the Japanese used detectives to scout out the genealogy of a potential partner, ensuring that any non-Japanese contribution was discovered before contracts were signed.

There was a Eugenic Exercise/Movement Association and in 1935 a Eugenic Marriage Popularization Society, a smaller part of the Japanese Association of Race Hygiene. Eugenic journals included *Jinsei-Der Mensch* ("Human Life"), and eugenic marriage questionnaires were distributed in magazines. As Popenoe encouraged Americans to collect and submit their genealogies for the greater good, so Japanese women were asked to assemble an account of all their relatives, so the information could, among other things, help prevent accidental inbreeding. In 1928 the twenty-first of December was declared "Blood purity" day, and free blood tests were offered to women.

In Germany the tangled threads of genealogy, heredity, and evolution developed along similar lines. Today people in the West tend to think of the impulses that led to World War II and the Holocaust as quite distinct from the character and preoccupations of people in the rest of the world. In fact there was much in common on both sides, especially as regards ancestry and its significance. Before World War II toxic beliefs about kinship, heredity, and race hygiene had led to mass sterilization of unwitting or unwilling people all over the world. In Germany the Nazi regime took the process to its horrifying extreme.

Chapter 4

The Reich Genealogical Authority

In due course, all *Volksgenosse* [racial comrades] will be placed in the position of having to show proof of their ancestry. For many racial comrades, it is of vital importance to be able to show this proof as quickly as possible.

—1939 German civil registrar's instructions, from
Eric Ehrenreich, *The Nazi Ancestral Proof*

When I met seventy-three-year-old Joe Mauch, he took a slim brown book out of his black briefcase and placed it on the table. The faded image of a gold eagle sat in the middle of its cover. At the top were the words *Deutsches Einheitsfamilienstammbuch*. It was a German family tree book.

The book listed Mauch's parents, Maria Lutz and Alfons Mauch and their parents and their forebears through to the late 1700s, Johann Michael Weider to Eleonora Weiss to Balthazar Luz. They were all *good Germans*, Mauch said, raising his eyebrows—they had good German names and they never traveled outside the country. The *Einheitsfamilienstammbuch* was a family tree, a document album, and a kind of passport for the entire family. It had slots for all the family births and marriages. Official stamps certified the marriage of Mauch's parents and the details of their parents' marriage. Mauch and his siblings—his brother, Jürgen, and their sister, Elisabeth, who died of diphtheria when she was three years old—had a page each, certifying when and where they were born. Mauch, born "Joachim" in 1940, had four official Stuttgart stamps on his page, one with a red horse rearing back, another depicting an eagle with wings spread wide above a swastika enclosed in a circle.

The book also contained a list of good German names for boys and girls: Joachim, Jobit, Julius, Jürgen. Essays at the front and back advised the citizens of 1930s Germany how to live productive lives. "Look," Mauch said, pointing at an essay title, *"Die Familie im Dienst der Rassenhygiene."* He said, "This word here, family in the service of *Rassen*, it means 'race,' and so 'race hygiene.' Keep it clean." We looked at another essay that explained why citizens should not marry people with genetic defects lest they pass them on to their children.

Mauch was open and thoughtful, and his conversation was punctuated by somber pauses and the occasional sweet grin that changed his whole face. His light blue eyes looked off in different directions; they both see, he explained, but his brain just won't use what his right eye looks at. He was born in Stuttgart on a street full of apartment blocks, and four years into the Second World War, when he was three, the street was bombed. "There were two signals," he said, "one a long, drawn-out siren, which means 'find shelter,' and one which went very quick, *dah, dah, dah, dah.* That meant that they were virtually arriving with their bombs."

He asked me if I'd seen the film *Slaughterhouse-Five*, based on Kurt Vonnegut's novel about the bombing of Dresden. Whoever created that film, he said, must have lived through it. "The bombing raid in the film, I nearly had to walk out of the cinema. It was very realistic, and it wasn't showing the fireworks. It showed a group of people going into this shelter underground, and then suddenly everything shakes, and then the mortar comes down, and there's a light globe hanging there, and then it starts to swing, and then it goes out. It's complete darkness. That was just how I remembered it."

On the day of the bombing, the apartment block next door to Mauch's was hit. A service tunnel in the basement of Mauch's apartment building linked it to the basement of the building next door, so after the bomb fell, Mauch recalled, they were dragging half-dead people through the tunnel into his basement. He paused. "What really got me upset was all the adults. My mother went hysterical and screamed, and for a kid I think that is not a good thing, the parents going to pieces." They were underground for probably half an hour, and when they came out, the apartment building on the right was completely gone, and Mauch's once-straight eyes were pointing in different directions. No one knew how it happened.

At the time, Mauch's brother had been sent to stay with a relative, and his father was on the Russian front. The only time Mauch met his father was when he returned home for his daughter's funeral. Some months later Mauch's mother, Maria, received a letter from the German government informing her that her husband was a hero, which meant that he was no longer alive. He died in a good cause, the letter explained. It did not say how.

Mauch grew to hate Germany. He and his friends grew up amid rubble and confusion, in a nation where eleven million people were murdered and many more incarcerated in death camps. But, Mauch said, "No one would tell us what happened." He kept asking questions, but no one gave him a good answer. "Everybody we queried, adults we called and talked to, they would always say they were goody-goodies, they had nothing to do with it." When his mother reluctantly told him something about the horrors visited on Jews by all the good Germans, Mauch would ask, "Why didn't you do something about it?" "Which of course is totally unfair," he said now, smiling sadly. His mother would say, "You wouldn't understand it."

"*No one* was a Nazi," he said. "*Everyone* knew that Hitler had been a criminal from the start." I asked him if his mother was a Nazi. "I don't think so," he said. "They had to be careful what they said in front of kids, because teachers would ask children to tell them if the parents said anything nasty about Hitler." Mauch's mother once told him about a day when the Führer made a televised speech. A neighbor came by afterward and asked accusingly if they had watched it. Mauch believed that his mother would not have told him that story if she was a Nazi herself, though she said her brother was. What about Mauch's father? "I don't think so," said Mauch. "He was a very strict Catholic, from what I've heard. That wouldn't probably go that well with being a Nazi."

By the time Mauch went to school, many pages in his biology textbooks had a blank sheet pasted over them. He was told later that these passages made the case for biological supremacy of the Aryan race. There was only one teacher, Mauch said, who "made it his job to teach us about what happened." Mauch remembers that the man was young and angry, although he cannot remember his name. One day the teacher brought in an architectural plan for one of the concentration camps. He showed the students the killing rooms and explained that in one room, where they shot people in the neck, a perfectly located channel on the floor was designed to drain away the blood after they fell. "Now just imagine the draftsman sitting there creating these things," the teacher told the children.

"I still can't comprehend how people could do that," Mauch said.

In 1960 Mauch was called up for service but instead of joining the army he fled Germany. He went to Australia but because he couldn't speak English he initially spent time only with other Germans. It was "the first time I actually met Nazis," he said. "They all agreed that Hitler was bad news and the Nazis did terrible things, but then they would find excuses and defend that behavior." So Mauch distanced himself from them and felt bad about being German for many years.

When he began to befriend the locals, Mauch was surprised to find that some Australians actively investigated their own genealogy. For him the pursuit had sinister undertones. Certainly his *Einheitsfamilienstammbuch* gave grist to the specific claims made by the critics of genealogy, as well as justifying their sense of panic. It wasn't just the rich or the aspiring who cared about family history—the Nazis did as well.

The *Einheitsfamilienstammbuch* was created in the 1920s by the Reich Federation of German Civil Registrars, and it soon became standard legal proof of genealogy throughout the country. According to Eric Ehrenreich, whose book *The Nazi Ancestral Proof* is the most specific and detailed account of the Nazi bureaucratization of genealogy and race, the registrars were quite explicit in their hope that it would be "a good means of advertising the goals of eugenics." Ehrenreich traced the origins of the *Einheitsfamilienstammbuch* and other documents like it to a period long before the Nazis came to power, when genealogists had significant social influence.

As early as 1898 the German historian Ottokar Lorenz described genealogy as a bridge between history and science. He argued that historians should think more about heredity, and scientists should think more about genealogy. Before Lorenz the science of genealogy consisted of record collection and organization, but at around this time genealogical societies began engaging with doctors, researchers, and psychiatrists who were studying heredity.

One of their first case studies was royal families, not because genealogists were slavishly drawn to power but for the obvious reason that aristocratic genealogies had all been worked out to a significant degree and because portraits existed of the principals by which at least some traits could be observed. Lorenz was professionally interested in the protruding jaw and lip that recurred over generations of the Habsburg family. He pointed out that the family's habit of marrying close relatives made the trait even more likely to appear across generations.

Family history was widely popular in Germany at the time, and in 1903 the *Zentralstelle für deutsche Personen- und Familiengeschichte*, the German Central Office for Family History, was formed. In 1908 the office made a formal commitment to collect genealogical information that would help psychiatrists and eugenicists understand "heredity, degeneration and regeneration." The goal was to gather the current and ancestral genealogical records of the entire population, from the aristocracy through to the bourgeoisie, and even including the inhabitants of prisons and asylums.

Here, as elsewhere, genealogy, heredity, and evolution were prime material in the grand exercise of science as public entertainment. The German Society for Racial Hygiene was founded in 1910 and in 1911 sponsored an exhibition in Dresden, a biological extravaganza that showed how cells worked and how hybrids were created. Family trees illustrated heredity, showing that musicality, "moral insanity," and, more straightforwardly, night blindness could be passed down. Reflecting the growing communication between genealogists and the medical community, genealogists gave talks at the exhibition. Such was the intense mingling of the disciplines that genealogical manuals began to include essays about the use of family history in psychiatry and anthropology. At the same time psychiatrists were engaged in a discussion about how to standardize records of the family histories of their patients.

As it did elsewhere, the combination of science and genealogy served as a way to unify the nation. Naturally, it wasn't long before concerns about heredity and traits became entwined with concerns about disability and race. Indeed, genealogy came to stand at the crux of historical, scientific, and nationalist interests. As far as race was concerned, as genealogists became more preoccupied with racial groups and scientists became more interested in how evolution shaped people, they looked at the differences between not just families but entire populations. What happens when a population is isolated and doesn't breed with others?

Scientists searched for island groups and groups who intermarried and reproduced only with one another. In many cases they ended up studying native peoples who were under the yoke of colonial rulers with whom there was already a great deal of tension. As one historian observed: "For Swiss anthropologists, the most obvious isolated group to study were the inhabitants of the Swiss Alps. For Americans, Native American populations were among the most promising isolates. For Indian and British scientists, isolation was to be studied in the caste system." One geneticist who was eager to explore racial purity and racial mixture said, "We have in the American Negro Population almost laboratory conditions for the study of the effects of racial crossing."

Anti-Semitism, in particular, was tightly linked to German genealogical activities, and as Ehrenreich writes, "The line between promoting the idea that distinct biological races existed and asserting that they were of differing value was extremely thin." In Germany Jews, who were thought to be both racially pure and foreign, were seen as a useful target population for genetic and evolutionary studies, much like Darwin's finches. Because so much conflict with and discrimination against Jews existed, scientists sought to explain social judgments with reference to heredity, not bigotry.

In the 1920s it became a common practice for genealogists and other interested parties to establish which families had Jewish blood and then to publish lists of their names. In 1925 there was a call to create a special eugenics division of the civil registrar's office that would record four to six generations of family history and biological information. From 1928 to 1932 eight volumes of *The Jewish Influence and the German Universities* were published by Achim Gercke, who headed the Genealogical Authority but later joined the university system. The volumes listed the names of faculty who were Jewish and part Jewish and even faculty who were married to Jews. At the same time genealogical journals increasingly called for measures like sterilization or worse to contain the threat of hereditary illness.

In retrospect it is easy to assume that ideas about racial hygiene became popular with the rise of the Nazis, but as Ehrenreich explains, "by the time the Nazis assumed power, virtually all of the basic components of their racist eugenic theory had already appeared in Weimar-era genealogical journals." Indeed, the ancestral proof that Nazis began to require of German citizens would not have been possible without the formalization of genealogy that began in imperial Germany and continued through the Weimar era.

Even before the Nazis were in government, party members had to demonstrate the purity of their Aryan blood. A party newspaper article argued: "Dogs and horses have family trees. Cattle are registered in herd books. This is the first condition to keeping the blood pure and will further yield a true Aryan foothold to the kinship group." When they came to power in 1933, they created an enormous administrative apparatus to classify the racial purity or mixture of sixty million German citizens, defining the populace's rights on that basis. "The interest in genealogy culminated under the Nazi regime when numerous eugenic databases were created," writes historian Bernd Gausemeier, "and the right to live became virtually dependent on one's family chart."

From the early 1930s the daily tensions and demands related to establishing proof of ancestry increased and even began to attract the attention of the foreign press. In 1934 a report stated that the Reich's minister of posts had directed all of his employees to produce evidence of Aryan descent. Earlier people had to prove they were Aryan only if there was doubt about their racial status. That same year Hitler's government decreed that only Aryans could hold stalls at the upcoming Leipzig Fair and that all wares had to be German made. There was no restriction, however, on "Jewish attendees or other Non-Aryans"; anyone who wanted to buy German wares was free to do so.

In 1935, in one of many articles about exclusions based on Aryan status, the *New York Times* reported that a young female clerk had been sentenced to four months in jail in Berlin for falsifying her grandfather's birth certificate, which showed that he was Jewish. The girl had erased "Jewish" and written "Evangelical" because it had become necessary to provide proof of ancestry at her workplace in order to keep her job. Some jobs required the absence of Jewish blood from 1800 on.

Nazi genealogy was not merely a way to bureaucratize the social categories that mattered to anti-Semites; it was an enormous social machine that reinforced, as well as recorded, the racism and eugenic ideals of the Nazis. Tens of millions of people were caught up in the documentary maelstrom. Each day Germans lined up at registrars and before other government representatives to produce documents establishing their ancestry. "Businesses gave away genealogical tables as marketing devices, much as present-day companies give away pens and calendars," writes Ehrenreich. Accordingly, many benefited from the Reich's preoccupation with ancestry, not just scientists and genealogists but also the gatekeepers of information, like civil registrars and churches. They saw their status rise and they profited through increased government funding and greater prestige. Genealogical magazines sold well, as did books with titles like *How Do I Find My Ancestors: A Guide to Quick Proof of One's Aryan Ancestry.*

For genealogists the new power was intoxicating. "For decades, kinship research was science's Cinderella," one wrote in 1936. "While other branches of learning were represented by university chairs, and encouraged by the state, people dismissed us with a pitying laugh. That has now changed thanks to the regime of Adolf Hitler. Today genealogy has tasks of state-level importance to fulfill."

In 1936 the Reich Federation of German Civil Registrars produced a new family passport, the *Ahnenpass.* Like the *Einheitsfamilienstammbuch,* it was a conveniently pocket-sized version of all of one's genealogical information, and once officially stamped, it functioned as a legal document. Many millions were produced; private companies created more than twenty competing versions. The Reich was so enthusiastic about the *Ahnenpass* that the military high command and even the

Office of the Führer's Secretary promoted its use. One featured a quote
from Hitler on its front page:

> There is only one most holy human right and that right is at the same
> time the holiest obligation that is to care that the blood remains pure
> and through the protection of the best of human kind the noble de-
> velopment of this essence to give the possibility of the development
> of this noble essence.

In practice, many Germans were able to "prove" their Aryan ancestry by virtue of taking an oath. Local administrators certified individuals at their discretion, and the assumption was always that further proof
would be provided once the war was over. Yet in the early part of the
twentieth century there were many marriages between Jews and Aryans,
and they all considered themselves German. Their children, who generally grew up to be Christian, thought of themselves as German too. Who
knows what steps the Third Reich might have taken had everyone truly
been forced to produce the details of his lineage?

Hundreds of thousands of individuals who were unable to gain approval with just an oath had to deal in some way with the Reich's Genealogical Authority. Ehrenreich combed through hundreds of wartime
letters written by ordinary Germans pleading with the authority for a
swift and favorable ruling. "No one knows my unbelievably heavy sorrow," wrote a woman whose son wished to marry an Aryan woman.
"Please leave me this little ray of hope," wrote another. "I grasp so desperately at your help."

The research was a surreal experience for Ehrenreich, whose mother
escaped Germany in 1939; his father survived the war but lost his two
younger sisters and many other members of his family to the Nazis. From
an early age Ehrenreich had been fascinated by the Holocaust: *Why
would someone want to kill all the Jews in the world?* The letters that he
read were housed in the German Federal Archives, the same site used
by Hitler's personal SS bodyguard during the war. During the day he
would come across letters from racial experts who had lived in the neighborhood and at night walk past the houses they had lived in.

By 1935 the Nazi government issued a law that prohibited marriage between the genetically healthy and "unhealthy." In 1939 Adolf Hitler created the secret T4 program, in which thousands of disabled people, along with the economically unfortunate, "burdensome lives," and "useless eaters," were reclassified as "life unworthy of life." In many ways it served as a pilot for the Nazi death camps.

Initially the program took in only children. Parents were encouraged to send their disabled offspring to special centers for treatment. Once there the children were starved to death or injected with a lethal overdose. As the program grew, people who were already institutionalized because of schizophrenia, epilepsy, dementia, or other disorders were transferred to one of six killing centers by SS soldiers in white coats. There they were led into chambers disguised as showers and gassed to death. Their bodies were burned in specially installed ovens, and relatives were notified of the deaths and sent falsified death certificates. Somehow news of the program became public, and even in Nazi Germany people objected. Eventually a grassroots campaign forced its closure, but it continued in secret. More than two hundred thousand people were killed in the T4 program.

When the war ended, Nazi genealogy and eugenics were finally put on trial. Hitler's personal physician, Major General Karl Brandt, was apprehended and prosecuted at Nuremberg. He created a program that sent sick, disabled, aged, and "non-German" people to the gas chamber to be killed, murdered people for the sole purpose of harvesting their skulls for medical research, and conducted medical experimentation where victims were sterilized, operated on, poisoned, or exposed to diseases like smallpox or terrible conditions like extremely high altitude. Brandt's defense included a copy of Madison Grant's *The Passing of the Great Race*. He specifically drew the court's attention to the parts of the book where Grant advocated activities that the Nazis ended up carrying out:

Mistaken regard for what are believed to be divine laws and a sentimental belief in the sanctity of human life tend to prevent both the elimination of defective infants and the sterilization of such adults as are themselves of no value to the community. The laws of nature require the obliteration of the unfit and human life is valuable only when it is of use to the community or race.

But Nazi eugenics did not just seek to eliminate the bad; they also sought to facilitate what they saw as the good.

Gisela Heidenreich's father died on the Russian front, and while it was not unusual among her school friends to not have a father, her friends had photos of theirs, and they knew their fathers' names. Heidenreich did not. Again, unlike her friends who were born locally in the small Bavarian town of Bad Tölz, Heidenreich was born in Norway. Her mother told her that in 1943, when she was pregnant, she had to work in a Lebensborn clinic in Oslo, which is where Heidenreich was born. Heidenreich had never heard the word "Lebensborn" before and assumed it was the name of the clinic.

Still, she always felt there was something that she hadn't been told. "You know, as a little child, there is something wrong, there is something weird, and you are feeling: *There is something I want to ask. I want to know and the answer is . . . oh, what? No, there is nothing, it's your imagination*," she told me.

When she was thirteen a scandalous story made headlines in Germany, and a friend quietly passed her an article about Lebensborn clinics. It claimed they were actually brothels for SS soldiers. Heinrich Himmler had created a breeding program using women who, depending on the story, were either prostitutes or innocent Aryan girls who were raped by soldiers (presumably with good Aryan documentation). *Oh, God*, thought Heidenreich. *Now I know why she never talked about my birth, why she never talked about my father.*

There was no one Heidenreich could ask about it. Her mother was a depressive, difficult woman who rarely laughed and shared little. She never even told Heidenreich about menstruation. "You cannot imagine how this society was. After the catastrophe of the so-called Third Reich, they were more and more withdrawn, even on the topic of sex," Heidenreich told me. "So I could not ask her, 'Is it true you were a whore?' I just had to accept that I was the product of having been bred in this brothel. It was horrible."

After she turned eighteen, Heidenreich answered the door one day to find a young girl standing there. Her visitor was the same height as Heidenreich, and she looked as if she was the same age. "Hello, Gisela," the girl said. "I am your sister." Heidenreich slammed the door shut but a minute later opened it again. Her new sister told her that they shared three other siblings and a father. Not only that, but their father was still alive and, most amazing of all, she said, "Our father is a beautiful father, a wonderful father, and he is very kind and loving."

It was like a fairy tale: Gisela loved her new siblings and her new father, who told her that he had always wanted to find her. Even her father's wife—the woman to whom he had been unfaithful—welcomed her and made her feel like a daughter. Heidenreich still lived with her mother, though they rarely discussed the situation. For the most part, when Heidenreich headed out to visit her father, her mother would say, "Say hello to him." When she was on her way back again, her father would say, "Say hello to her."

Heidenreich's father told her that almost twenty years earlier he had had an affair with her mother. It was a love affair, and yet he also told her, "It was an order from Himmler for his SS men to produce children out of the marriage to pass on their precious blood." Not wanting to give up the happiness she had so longed for, Heidenreich didn't ask for any other details, nor did she ask her father what he had done in the war. No one ever mentioned it.

In fact the founding principle of the Nazi Lebensborn program was to ensure that no Aryan children were aborted. Abortion was illegal, and doctors who performed the operation were executed. If a woman found herself pregnant outside of marriage, and if she could prove that both she and the father were Aryan dating back to at least 1800, she could give birth to her child in a Lebensborn clinic in secret. If she wished to leave her child with the program, it would be adopted out to an SS family or raised in a Lebensborn home. The clinics were not actually brothels or places of rape, as the lurid coverage from the 1950s had suggested, but rather were deluxe destinations for the wives of SS officers to give birth to their children. SS soldiers paid for the clinics out of their salary.

Still, Himmler's zeal to create a master race included not just providing sanctuary for unwed mothers but actively encouraging SS soldiers to father more children. Eight thousand Lebensborn children were born in the homes. Heidenreich told me that about half of them were taken home by their mothers, and half were left for adoption. In Nazi-occupied foreign territories, children who looked Aryan were kidnapped and deposited at Lebensborn homes to be raised as Germans. It's thought that twelve thousand children were born or abducted in Norway. Up to one hundred thousand were taken from Poland, and it may be that overall more than two hundred thousand children were removed from their parents in the Eastern Bloc countries to be Aryanized. Documents show that some parents signed their children away, but this took place in occupied territory so it's doubtful how voluntary their choice was.

After the war the Lebensborn clinics were investigated by the Nuremberg trials, but it was concluded that they were charitable institutions. Heidenreich's mother was a witness in the trials. "She lied," Heidenreich told me. "She testified that the clinics were merely places to help women give birth, but they committed many crimes against humanity," she recalled. "She always said she had been just a secretary, but that was not true." Heidenreich's mother was a senior member of the staff and managed the identity change and dispersal of stolen Norwegian children. "I think she even spoke Norwegian," Heidenreich said, "but she never confessed."

The fate of the Lebensborn children was a miserable one. Many were abandoned by their mothers and brought up in orphanages—most of which were terrible places in which to grow up. Gudrun Sarkar, who is now seventy-three years old, was left in a Lebensborn home until she was eight years old and still suffers phobias because of it. The Lebensborn nurses were so particular about how children ate, Sarkar told me, they insisted that children wear a bib when eating and that half of it cover their top and the other half be stiffly lodged under the plate to catch any food. It was such a tense balancing act that Sarkar still has trouble eating or sitting in a dark room. In elementary school Sarkar discovered that she was a Lebensborn child when her teacher asked her where she was born. When she told him it was a Lebensborn clinic, he explained that she was born for Hitler, and she should be ashamed of herself.

Eventually Sarkar was adopted by elderly Germans who remained committed Nazis long after the war ended. They believed it was their duty to raise an Aryan child, but they were not kind to Sarkar. Hitler was good for Germany, they said, but she was born in shameful circumstances. When she was a teenager, Sarkar brought home photos of dying Jews in concentration camps, but her parents said they must be fabricated.

Few Lebensborn children were able to find their original families, and many were rejected when they did attempt to make contact. As an adult Sarkar was discovered by her mother's sister, who found her only after her mother died. They had a loving relationship. She also found out where her father, a member of the SS, lived, and she learned that she had three half-brothers. She never got in touch for fear they would not want to know her.

In Norway one doctor declared that any child of the SS must be mentally defective, so he placed them in mental asylums, only to be released when they were in their twenties. Even when children were kept by their mothers, local communities often stigmatized them. Heidenreich was unusual for having found a father who wanted to claim her and an extended family who embraced her. Still, as a young woman she was almost six feet tall, with blonde hair and blue eyes, and she was deeply embarrassed to so obviously embody the ideal of the Reich. For a long time she dyed her hair dark brown.

Heidenreich became a special-education teacher who worked with disabled children. She believes now that unconsciously she wanted to help children who would have been exterminated by Hitler. Later she became a family therapist specializing in family-systems therapy, a method that explicitly encourages exploration of the past. Issues that are unresolved in previous generations, Heidenreich advised her clients, have a way of surfacing in the present one. It was fifty years after the terrible events that took place when she was thirteen before Heidenreich realized she should take her own advice.

In fact, documents relating to the program and containing the names of parents were released to the public only in the early twenty-first century. It was then that most Lebensborn children became aware of the circumstances of their conception and birth. A support group was started in 2006, and some individual Lebensborn children have begun to speak publicly about their experiences. Allegedly some are proud to be part of an elite group, but there is little evidence of special status. In 2006 Lebensborn child Ruthild Gorgass told the *New York Times*, "My eyes aren't perfect. We've got all the same illnesses and disabilities as other people have."

Heidenreich went through SS records and found that her father ranked high in the organization, but he was posted in communications and never accused of a crime against humanity. She discovered that her mother was a Nazi and read transcripts from her testimony at the Nuremberg trials. "She always told me that she brought me back with a transport of other babies to Berlin," said Heidenreich. "I never knew what it meant, but in the end of course I found out this was one of the transports of Norwegian children." In fact, Heidenreich's mother first told her own family that Gisela was a Norwegian orphan. Later she acknowledged that she was her own child.

"After all the shocks and shame, we also feel guilt, which is funny," said Heidenreich. "But most of us feel guilty . . . [because we were] the product for the Aryan race, for this madness of this regime." Still, Heidenreich said she has turned her guilt into responsibility. Now she gives talks to schools throughout Germany, and she marvels at how little even the teachers now know about this side of Nazi eugenics.

Joe Mauch has lived in Australia for more than four decades. Now a slim and fit septuagenarian, he spent most of his adult life working with books, apart from eleven years when he tended his own olive grove. He has visited Germany a few times, and it was on one of his trips that his mother gave him the *Einheitsfamilienstammbuch*. He told me the pendulum had finally swung back—all children in Germany are now educated about what happened during the war. Mauch's brother, a professor who stayed, told Mauch there was even a school whose motto was "No more

Auschwitz." "It's pretty brutal," Mauch said. "I feel sorry for these poor little kids who walk through that gate."

Genealogy doesn't worry him anymore either. "I can understand that people are interested in their ancestors and where they come from," he said. "Genealogy is just like doing history. It's nothing to be sneered at."

It's hard to reconcile the criticism that family histories aren't worthy of serious attention with the experiences of Joe Mauch and Gisela Heidenreich and their peers, for whom big history was very personal. This paradox will only increase. The more we digitize records and make them accessible, the more we learn how to read DNA, then the more we will see extraordinary, fine-grained intersections between world history and personal history in the lives of ordinary individuals—not just royalty, not just the bourgeoisie, and not just Cleopatra.

Yet for all the people who took the journey through time that Mauch and Heidenreich did, there is still a strong current against the notion that examining the history of the people who formed us may help us learn about ourselves. Even as Western society devises ever-new ways to think about itself and help us understand "who we really are," and even despite the millions of people who subscribe to services like Ancestry .com and FindMyPast, curiosity about lineage outside the ranks of committed genealogists isn't taken very seriously. But history—if we care to examine it—shows us that genealogy is potent. Simply asserting that it shouldn't be is meaningless.

No doubt many people today have inherited the fear that genealogy is the first step down the path to eugenics, without ever quite knowing the basis of that fear. But awareness of one's role in the great historical narrative does not necessarily lead to delusion or bigotry. Nor does curiosity about personal history mean someone wants to be a queen. Rather, it is delusion and bigotry that lead to the misuse of records and ideas. The misuse then creates a fear of those records and ideas.

In the 1950s Friedrich von Klocke, who was a genealogist during the Third Reich, came to rue the role he and his associates played in it. He thought it had happened because he and his colleagues tried to make a science out of a field that was not scientific. But Eric Ehrenreich argues that genealogists like Klocke were merely doing what everyone else at the time was doing: embracing a racist ideology.

Still, the dogma that people shouldn't ask how their ancestors shaped them will not prevent a twenty-first-century eugenics movement. Indeed, the insistence that we shouldn't know these facts or try to analyze them or have feelings about them doesn't mean that the details of our personal history won't be interesting to those who may yet choose to investigate them for the wrong reasons.

Most recently, during the financial crisis in Greece in 2012, as the fascist Golden Dawn political party became more and more powerful, the question of how long a person had been in Greece and whether he could prove it began to be raised. In a *New York Times* article, a local cracked a joke about needing to show that you were Greek for three generations back. While some people will remain personally indifferent to their forebears—and there's no reason to argue that they shouldn't—it would be wise to be vigilant about how information about families is kept or lost or found or shaped by powerful socioeconomic and cultural forces over time.

The bureaucratization of genealogy in Nazi Germany was a suffocating expression of totalitarianism. Yet as genealogy can be used by totalitarian forces to persecute people, so can antigenealogy. Some regimes, whether institutional or political, control people by targeting their personal histories, not by using them against them but by taking them away altogether.

Part II

What Is Passed Down?

Chapter 5

Silence

History is important. If you don't know history it is as if you were
born yesterday. And if you were born yesterday, anybody up there in
a position of power can tell you anything, and you have no way of
checking up on it.

—Howard Zinn

In 1937, at the age of fourteen months, Geoff Meyer appeared before a
magistrate, who made him a ward of the state and handed him over to
a state-run orphanage. By the time he was four, Meyer lived in a "boys'
depot," which housed thirty to fifty children until they were fostered,
though many were fostered, returned, and fostered again. For all the
years he lived there, Meyer never learned any of the other boys' names.
"We weren't allowed to talk to each other," he said, "and the staff always
said, 'Hey you' or used terrible words."

Every day at the depot began with a reckoning for children who wet
their beds. Staff draped the urine-soaked sheets around bed wetters'
heads and made them parade around the dormitory. The other boys
laughed, Meyer said, until it happened to them. "I was too small to laugh
at anyone," he told me. "I was scared of being bashed up." The food was
often rotten, and when Meyer threw up after eating weevil-ridden por-
ridge, he was forced to eat his own vomit. Punishments included floggings and scrubbing the floor with a toothbrush, but the most feared was
the small cupboard under the stairs. Boys were locked in with no food or
water, and they soiled themselves until they were released. They never
spoke, Meyer said. "We held hands."

When prospective foster parents visited, the boys were lined up along the front veranda to be inspected. Meyer was fostered out eight times, and in his final placement he was sent to an old woman at Wentworthville, New South Wales. Meyer didn't know who his parents were, why he was a ward, or if he had any family members at all, and—like every other adult in his life—his new foster parent wouldn't tell him anything. It was only when he overheard her enrolling him at the local school that he found out his birthday for the first time. Still, everyone at the school knew Meyer was a ward because the assistant principal made him and another boy stand up and announced to the class, "They are under child welfare because their mothers never loved them."

On May 10, 1954, his eighteenth birthday, Meyer simply fled, never to return. He had twenty-four pounds and eighteen shillings, the clothes he was wearing, a tennis racket, a cricket bat, and no friends, acquaintances, or family that he knew of. He had no idea how to find a job or a place to stay.

In Australia, where Meyer grew up, at least half a million children were placed in institutional care in the last century. In the United States today there are more than thirty thousand children in such circumstances; and in Africa, Asia, and Latin America the official figures are in the tens of thousands, although it's believed that there are many more unofficial cases. For a long time these children were ignored, but in the last twenty years many stories about their mistreatment in homes and its long, damaging aftermath have emerged. All over the Western world, adults who were once institutionalized children have recounted vividly similar experiences of floggings, forced labor, sexual abuse, and emotional torment. In some homes children were not allowed to look one another in the eye.

The long-term consequences of such treatment have been the same the world over too. When the children reached a certain age and were ejected from their "homes," they entered their own country like refugees, knowing nothing about "councils or libraries or voting." Many died from drug- and alcohol-related causes, and some built careers in institutions, like the navy, or in the religious orders that ran their homes. Some became successful, but many struggled. Figures suggest that one in three attempted suicide, many experienced homelessness, and they had a high incidence of mental illness and physical injuries. Most "ex-orphans" were noticeably short (a trait usually attributed to malnutrition), and while they lived in terror of being forced into old people's homes, a number of their own children ended up in care. There is also a well-worn path from children's homes to jail: The last three people to be hanged in Australia grew up in such places.

Despite the fact that most people today know that terrible things sometimes happened to children in group homes, few comprehend that these institutions operated like totalitarian states within a democracy. As well as being places of mental and physical torture, the institutions systematically controlled children's access to information while they were institutionalized as well as once they left. In many homes the staff had authority over every connection children had with the outside world, not sharing news and even confiscating letters from family. Some children were even schooled at the institution and did not leave the grounds for years. Many were not taught to read or write or do basic mathematics. Their names were arbitrarily changed, and it was common for them to be addressed only by a number. Some children were told that their parents were dead when they weren't or that they never wanted to see them again when they did.

Even though it has been decades since group homes closed, vital information about their residents is still being withheld. For the children it's as if someone pushed them through the looking glass, and decades later they still can't find their way back.

What gets passed down? Records, of course, by definition are one key source of personal data. It may be a banal observation, but sometimes you don't notice what gets passed down until it doesn't. Birth certificates, school records, the names of family members, and all the other bits of information that we take for granted only become disturbingly obvious in their absence. The ordinary records that chart the passage of a life matter a great deal—not just to governments and corporations and librarians but to ordinary people too.

One reason why it's so hard to understand the sense of rootlessness experienced by the former residents of the group homes is that most people live their lives firmly embedded in a web of information. They know where they were born. They know whether their parents loved each other. They know what it looks like when an adult brushes her teeth. They belong to interconnected groups, whether a family, a neighborhood, or a religion, and they have experiences that constantly reinforce what they know. Together these bits and the threads that bind them add up to an incalculably crucial body of information, providing not only a history but also a sense of self. It's almost impossible for most of us to imagine not knowing such facts about ourselves, and yet this information was effectively erased from the lives of the children of twentieth-century orphanages.

One "ex-orphan" told me that these former group-home children want, like any other citizen, to have information about themselves and their family—or its proxy—and all the power that such knowledge brings. Yet there are enormous obstacles to obtaining it. In Australia records are scattered throughout each state, held by government records offices and by the religious institutions that housed children. Government departments may take years to respond to a single request. Many records were destroyed, but there's little clarity about what was lost and what was never kept in the first place. Many files are undated, sloppy, or incorrect, and there is no consistency in how files are searched for or delivered. There's no central organizing body, and most people need professional-grade archivist skills to find and understand the documents. Generally the ex-orphans distrust bureaucracy, and while it is intimidating enough for them to enter a neutral institution like a public records office, many must return to the actual organization that mistreated them. The overzealous application of privacy laws also means that when many care leavers do manage to receive files, their missing siblings' names are redacted. One ex-ward received a photo of a children's party with all the little faces at the table whited out except for his own.

In forty-eight of the United States most citizens are not only given automatic access to their original birth certificates but also protected by privacy laws so no one else can see them. But this doesn't apply to children who are adopted. At adoption their birth records are sealed, and a new certificate is issued with the names of the adoptive parents. Unless the birth parents have explicitly signaled consent for the adoptee to contact them later, adoptees in many states cannot gain access to their records without paying hundreds of dollars and getting a court order. Even then, adult adoptees who petition for access may be denied, and if they are successful, many states differ in what information they will provide.

In Texas and South Carolina adoptees are legally required to have counseling to deal with the possible emotional consequences of learning who their birth parents are. In Connecticut it is against the law to release any information that might assist in the identification of birth parents if a third party (such as the agency that initially placed the child) feels it would be disruptive to the adoptee or to the birth parents. In many states it is possible for birth parents to block the eventual opening of the record. In Minnesota in 2008, twelve hundred adoptees were unable to gain access to their basic birth information because of an affidavit filed by their biological parents. After all the social change of the 1960s, after the gains of the civil-rights era, after laws were passed making it illegal to discriminate against women and people of color, adoptees are the last group of American-born citizens who are denied straightforward access to this fundamental information.

Naturally, when information can be retrieved, even the most banal detail of a stolen life can be traumatic. "People get rotten drunk in order to read their files," one activist told me. In a government report one woman described opening her files at home alone and being committed to a psychiatric ward a week later. Others put the files away and never look at them.

Ivy Getchell went searching for information about herself at state government offices in 2004. Getchell was taken from her family by social services as a young girl and spent years inside the Parramatta Girls' Training School, an old convict prison, where leg irons and wristbands were still attached to the walls. While she was at Parramatta, Getchell's "name" was "Fifty-five." It wasn't until she was seventy-one years old that she felt able to start searching for the records of her life. Her father was long dead, but she found a file of letters from him that she hadn't known existed. He wrote:

> Ivy, my little mate, for Christ's sake answer my letters. Let me know where you are. I will come and bring you home. We miss you and love you. We have a nice house now up at old Kelly's place near Mount Bathhurst. You will remember it. Ivy, I have a job. I can help you. Please let me know where you are.

I met Geoff Meyer in 2012. With his Fair Isle sweater and slicked-down hair, he looked like anyone's seventy-six-year-old grandfather. He was courtly and jokey, and he called me "mate." He said that not long after he had run away to Sydney, "I started to get it into my brain to find out if I had any family." He guessed that the best place to look was the Department of Child Welfare. "I'm a state ward," he told a young man at the local office. "I'm looking to see if I've got a mother and father." The young man went into another room and after five minutes returned and said, "I think you might have a sister." He disappeared again to check further, and then an older man came out and said to Meyer, "I think you had better leave." Meyer thought he had misunderstood. "I think you had better leave," the man repeated. "No," said Meyer. They argued back and forth, the man continuing to try to dismiss Meyer with no explanation and Meyer refusing to budge. Then the man told him, "Get out, or I'll call the fucking police." Meyer was frightened of being sent back to his foster mother, so he left.

He got a job, married, and had four children and, as the years passed, eleven grandchildren. Meyer never told any of them that he had been a state ward. When his own children asked him about his childhood, he changed the subject. But when he retired, he started to go to the state records offices to see what he could find. Even then he didn't tell his wife about his past. "It felt very, very private," he explained. He eventually tracked down his birth certificate and discovered that his mother was Maisie Aileen Meyer, a Sydney local, and his father was Leo Joseph Meyer, an American sailor. There was no information about why he had been made a state ward and no record of contact from his parents. As he continued his search for records about his life and family, Meyer was told different things by different departments. Some officials were kind to him, while others were perfunctory. One said his information had been lost in a flood; another claimed it had been destroyed in a fire. At the records offices Meyer had to insist that he was legally entitled to a copy of his files. When he received them, they took months to reach him and were often missing documents from the original sets he had seen.

When he was sixty-eight, Meyer saw a newspaper notice seeking former state wards. He responded and shortly after found himself at the head office of the Care Leavers Australia Network in Sydney, speaking with one of its founders, Leonie Sheedy. "She started talking to me, and I talked, and then the more I talked the more she was getting out of me, and I had never talked like this before." When Meyer left Sheedy that day, he said, "I felt like Superman walking in the air. I felt like Jesus Christ walking on the water." That conversation reframed his life. "I thought I was the reason all that stuff happened," he explained. "All that time, I thought it was only happening to me, but it was happening all over the place." When he got home that day and told his wife about his experience, she asked, "What really went on in your life?" So he began to tell her too.

Over a cup of tea Meyer showed me the files he had recovered. The first item was his intermediate school certificate. He found other records about his education and his foster parents, but he could locate only one from before he was ten. He spends a lot of time now searching for the missing records of his first decade; they are proof of an otherwise invisible life, but he also wants them so he can sue the government for compensation.

The likelihood of Meyer's finding his files is remote. All over the world the funds devoted to preserving records and archives are being cut. Of course, the fewer funds that are available, the more endangered are the records that do remain.

In 2012 an Australian state ombudsman revealed that his local government had committed hundreds of breaches of records-management legislation. He discovered that a single department was in possession of eighty linear kilometers of children's home files, most of them uncataloged. Some were stored in basements with dripping water and rat infestations; others were illegally marked for destruction.

While freedom-of-information acts in most Western English-speaking countries may grant people like Geoff Meyer access to their records, information that has not been indexed can never truly be free. The care leavers suspect the governments are suppressing files to avoid being sued. Perhaps, but bureaucratic apathy achieves the same ends.

The last time I saw Meyer, he told me that he had discovered a distant cousin on his father's side in the United States through Ancestry .com. His new cousin told him that his father had returned to the United States and died young. Later Meyer's mother and another man also set sail for America. Their ship berthed in California, but there the trail went cold.

Meyer wondered for a long time whether he actually had a sister but never found a trace of her. He now believes the young man at the records office got the spelling of his name wrong. Meyer told me he'd had three heart attacks but he always woke up happy, thinking, *Another day!* He would keep looking for his files. What he wanted to know most of all was whether he'd been surrendered or taken. If he'd been surrendered, he reasoned, maybe the person who'd turned him in was a relative. Maybe it was a sister of his mother, maybe she had children too, and maybe he had more family. Still, he said, "I'm seventy-six. How long do I have to find out?"

Totalitarian power thrives when it alienates people from basic information about themselves. When European slavers abducted people from Africa, they essentially took away their history as well. It was a profoundly dehumanizing act that occurred because the system treated the individuals who were caught up in it as less than human. In Canada and Australia in the early to midtwentieth century, many indigenous children were taken from their communities and raised in settlers' families or group homes. This act has since been described not just as abduction but as cultural genocide.

Other regimes have specifically targeted historical and family information for destruction. In 1924, when communists were in power in Mongolia, they destroyed family trees that had been kept for generations and banned indigenous surnames; for more than seventy years locals simply called one another by their forenames. In 1998 the government decreed that Mongolia's citizens must rediscover and register their surnames and their fathers' names. But by that point many people no longer remembered them.

In North Korea today in the *songbun* system of social organization, party and local administration officers track the ancestry of all citizens, but they don't share that information with the country's people, who may well have lost personal knowledge of it. A family connection to a dissident—be it a close relative or a distant ancestor—will block a person's access to some jobs, to education, to party membership, and even to food.

In Mao Tse-tung's China, the leader's view of genealogy was initially benign, until he began to feel threatened by the established power of traditional clans. Baiying Borjigin, a Beijing native of Mongolian descent, began to explore his family history at the beginning of the twenty-first century, eventually writing a book about his quest, *Searching for My Source*. As a child in Beijing, Borjigin was told that his family was descended from Borjigin Temujin, otherwise known as Genghis Khan. Indeed, Borjigin was the name of the clan into which Genghis Khan was born. Yet before he could investigate his connection to the ancient warrior, Borjigin had to deal with the fallout from the recent dictator.

I met with Borjigin, who is now in his seventies. He told me that his family entered China many generations before he was born, and most of his Chinese ancestors were buried in a long-established family graveyard in a beautiful forest outside Beijing. The last members of his family to be buried there were his father's parents, who contracted tuberculosis. At that time, Borjigin wrote, TB was seen as a terrifying pestilence, so his grandparents were sent to live out their last days in a residence inside the family graveyard. There they were looked after by a peasant family that was paid to maintain the cemetery. Borjigin's father was sent to stay with relatives.

When Borjigin was six, the Chinese government reclaimed the land of his family graveyard, and his family was told to remove the bodies of its ancestors. One memorable night Borjigin watched his father and other relatives bring many strange clay-covered objects into the house. He peered from behind a door to find them washing the clay from objects made of gold and silver with emeralds and other precious materials. When he asked his father about it, he was told, "This is no business for children!"

Later Borjigin's great-aunt took him to the public cemetery where the remains of his ancestors had been reinterred, now without their funerary treasures. She told him that it was his birthright to be the keeper of the family's memory. "You are the first man-child of your family, and you are very obedient. Everyone else can forget our family grave, but you can't," she said. "You must often add some earth to our family graves when you grow up."

But the message changed again in the mid-1960s. Like many other Chinese, Borjigin's family was forced to turn its back on its history with the onset of Mao Tse-tung's Cultural Revolution. By now the family fortunes were so reduced that they were regarded as urban poor, yet in the eyes of the regime their background relegated them to the "exploiting classes."

During the Cultural Revolution, young people all over the country banded together in Red Guard groups. They harassed and terrified the citizenry, invading their houses and looking for evidence of membership in the malcontent bourgeoisie. If the Red Guard found *jai pu* (ancient family history records) or other evidence of ancestral power, they would destroy it on the spot and quite possibly kill the residents of the house as well. Borjigin's family no longer swept its family tombs or looked after the graves, as even those behaviors were regarded as "evidence of the restoration of reactionary life." The worship of ancestors and other old customs were forbidden. "It was very dangerous," Borjigin explained to me through a Chinese interpreter. "Family had to destroy itself."

Some families buried their *jai pu*, others hid them in walls, and some destroyed them before the Red Guard could. One day, as the Red Guards searched houses down the street from Borjigin's, his father told him to find any documents that might reveal their family history. They gathered the deed to the family graveyard and the map of its graves, which held more than forty ancestors from over ten generations. Each ancestor was accompanied by a painted portrait and a description of his or her position in the family tree. Dozens were officials in the Qing dynasty. Just one of those paintings, Borjigin knew, might lead the Red Guard to beat his entire family to death. He and his father found a large basin, and Borjigin placed all the documents inside it and burned them. His great-aunt and his father forbade him to ever mention the family graveyard or his family history again.

Borjigin was fifty years old before he was told the names of his grandparents. In 2000 he went looking for the family plots he had seen decades before. There had been a great deal of urban encroachment in the area, yet the site that Borjigin had visited as a child still looked familiar. The graveyard had, however, since been turned into a vegetable plot for the nearby university, which employed peasants to farm it. Borjigin's companion asked the peasants if they knew the plot had once been a graveyard. They said it had been a vegetable plot as far back as they could remember, although they did often come across human bones when they were digging there.

Standing where the graves were once marked, Borjigin was sad but also embarrassed about his sentimentality. Long before his family had become wealthy and then poor again in China, they had lived and died in the deserts of Mongolia. Their bodies went into the sand where they died, and they never expected their descendants to worship them, he thought. Now his ancestors were fertilizer in this soil. "The meals and bodies of the elite of modern China, the teachers and students of the national university, were improved. This was the best final result, wasn't it?"

Borjigin told his father before he died that he was going to reconstruct the family tree. His father replied that he had never regretted selling their jade and antiques and paintings but that he thought every day about the family tree they had burned. "If you can restore this list," he told Borjigin, "I will rest peacefully when I die."

The current Chinese government has supported its people's reverence for their heritage. China's largest libraries, the National Library of China and the Shanghai Library, have hundreds of thousands of family genealogies. "The clans are back in full force," one researcher told me. "Some of them have two to three hundred thousand members and they are fairly flush with money and all about preserving the family." Now the government doesn't persecute people for their family graves. It charges them rent instead.

There has been a similar resurgence of interest in family history in Eastern Europe. Some researchers suggest it's a reaction to the void left by the regimes that were overthrown in the 1990s. "They yearn to reconnect to their family, to their roots," one observed.

Within families it is, of course, the adults who seek and manage this information. How they choose to share it with the smallest offshoots in the family tree is a private matter. Certainly the control that adults exert over the information in the lives of children can be extreme, and sometimes the regime that withholds information is not one's government but one's own family.

"Who has a convict in their family tree?" asked my eighth-grade teacher. Student after student held up a hand, testifying to their awesome convict heritage, while I sat quietly resenting them. Having not learned my lesson when I was seven about posing certain questions, I later asked my father if we had any convicts in the family. I reasoned that I had never been actually told there were *no* convicts in our line, as the question had never come up, so maybe all I had to do was ask, and it would be revealed. I was wrong. My father, who was unmoved by the teenage notion that it was now cool to be of convict stock, dismissed the idea.

It was therefore interesting a few decades later to learn that the woman who raised my father—his grandmother—was in fact the daughter of a convict. The discovery came about when I asked a local historian to help me find my family's origins. In a surprisingly short amount of time she led me to documents about my great-great-grandfather, Michael Deegan, about whom I had never heard a thing. The convict register in the Tasmanian Archives recorded that he was sent to Port Arthur in Van Diemen's Land in 1842. I could hardly believe it—this was the Australian genealogical jackpot. For one small and deeply satisfying moment, I was a convict princess.

In the clean, crisp air of Tasmania today, the remains of the peniten-
tiary at Port Arthur look more like an elegant old manor than a jail. Even
though a fire rendered the main building unusable in the 1890s, the
four-story goliath still dominates the scene. Standing beside a deep bay
and surrounded by green hills and a forest of blue-gray gum trees, it is
the largest building in a complex that includes a church, an asylum, a
hospital, and a manicured garden, all surrounded by English oaks. Farther up the grassy slopes, the commandant and his colleagues lived with
their families in a row of pretty Victorian houses. Out in the middle of
the bay on the tip of a peninsula was Point Puer, a rehabilitation institution for younger criminals kept separate from the main group of convicts.
Between the point and the main site is the Isle of the Dead, where more
than 1,500 people were buried.

The black heart of Port Arthur was the Separate Prison building,
where warders forswore whipping and tried to more directly manhandle
the souls of their charges. When convicts arrived there, they were
hooded so they could not look at or speak to another soul for all the years
they were incarcerated. In the prison chapel wooden doors sealed all
men off from one another. Each seat, or rather each place to stand, was a
narrow, suffocating box that was constructed so the only thing the convicts could see was the preacher. Parts of the building were based on a
panopticon design, which permitted a guard standing in place to survey
four tiny yards where prisoners walked in a small, quiet circuit. Prisoners who misbehaved—or who were made so mentally ill they could not
contain themselves—spent time even more completely isolated in a
stone cell that let in almost no light, even on a bright day. Outside the
Separate Prison, convicts labored in the pastoral loveliness to produce
bricks, worked stone, ships, and furniture. They wore uniforms of
worsted wool, of different colors depending on their status. One jacket,
known as the magpie, was specifically designed for humiliation: Its panels, sleeves, and each side of the collar alternated black and yellow, and
it made the convicts look like jesters.

Young Deegan had been transported to Van Diemen's Land after
having been arrested three times in his hometown of Dublin, at which
point he was sent to Point Puer and incarcerated with boys as young as

nine. When he disembarked, he was fifteen years old and five feet and three quarters of an inch tall, with hazel eyes and an inoculation mark on his left arm. He left behind two parents and a brother, and it's likely that he never saw his family again.

When I went through Deegan's records with a historian, she noted the surgeon's comment about his "troublesome" conduct onboard the *Kinnear*. Boys on transport ships were at risk of rape, she said, and fending that off sometimes made them aggressive. They were often hungry too. Yet while Australia's convict system left a huge number of men and women so traumatized they couldn't function, many convicts were given a chance to change their fate. "The British state invested in them by teaching literacy and skills to the young ones," the historian told me. She looked at Deegan's files again and pointed out that when he arrived at Point Puer he couldn't read or write, and he only had one offense for all the time he was in jail. "See," she said, "he learned how to be good."

After Deegan moved to Victoria, he married a woman named Ann McGrath. They lived in the countryside of Castlemaine and had ten children. Did McGrath know that her husband had been a convict? I have not yet been able to find out. After all, it is a challenge to look for information that has gone missing at some point in the last century and a half. It also made me wonder how silence is passed down.

Until the 1970s there were no convicts in Alison Alexander's family tree. That changed when a distant relative of Alexander's father, a Mormon, began digging into the family's past. He found that the family descended from Jane Baird, a Scottish woman who was sent to Van Diemen's Land for robbery with violence. Alexander remembers that her father was delighted to learn about Baird, and when he discovered there were more convicts in the family, he had a tree made up with little pictures of all the items they had stolen: a silver watch, two pairs of stockings, and ten silver cups. Alexander's mother was less pleased with the discovery, and her own mother, who was in her seventies at the time, was appalled. She couldn't understand why her son-in-law would want to know such things or why he would then take pleasure in what he found.

"Never mind, dear," she assured Alexander. "There's nothing like that on our side of the family."

But when Alexander, who became a historian and a writer, began to do some research herself, she discovered at least nine convicts in the family tree. Two were sent to Australia on the first fleet in 1788, and they married other convicts and had children who grew up and married the children of other convicts. Alexander ultimately determined that three of her four grandparents had descended from convicts. After further research, she found that her husband was the first person to marry into her family who hadn't descended from a convict.

On the face of it, Alexander's new family history made perfect sense. She and her parents were born in Tasmania, and Tasmania was once the world's largest penal colony, with more criminals per capita than the next two candidates, New South Wales (another Australian convict colony) and Siberia. In fact, by 1853 fifty-one thousand of the total sixty-five thousand Tasmanians were convicts or ex-convicts. Because there hasn't been a lot of immigration to Tasmania since that time, it seems only reasonable that Alexander would have at least one convict ancestor. So why didn't she know that was the case?

If it was odd for a Tasmanian historian to have no idea about her family's past, the experience was almost universal. In 2009 it was estimated that three quarters of the current Tasmanian population had convict roots. Yet hardly anyone seemed to be aware of his lineage. The oldest Tasmanians that Alexander interviewed were born in the 1920s, and when she told them that most Tasmanians were descended from convicts, they were utterly astonished. Although the convict system ended in the 1870s, another historian told me that as early as the 1920s people simply stopped being aware of who had come from convicts and who hadn't. Even at the beginning of the twenty-first century, Alexander reported, only 36 percent of Tasmanians said they knew of a convict in the family tree.

I drove from Port Arthur to another tiny Tasmanian peninsula to meet with Alexander, now one of the world's experts on the island's convicts. Her house made me think of nothing so much as freedom. From her front and back windows you can see the sea, and if you flew due south from there, you'd be in Antarctica in a mere four or five hours. How, I asked her, had an entire island come to forget its heritage?

In the eighteenth and nineteenth centuries, the British hung the worst of their criminals; they transported the lesser offenders to Australia. One contemporary observer said that it was the first time that people had tried to create an entire society from all that was wicked in another.

Van Diemen's Land received its first shipload of convicts in 1812, and originally the settlement was a curiously laid-back place. Far from the Old World, the small population had to rely on one another to survive, and the distinction between convicts and captors became moot. There wasn't a lot of law and order in the penal colony, and corruption was common. Although punishments were meted out to lawbreakers, bribes could be paid to circumvent them. Couples shacked up together and moved in and out of relationships without feeling obliged to get married or even appear respectable. Sadly, the lenient founding atmosphere didn't last. Soon new representatives of the Crown arrived, bringing more convicts and a more severe Victorian morality with them.

Still, Australia remained a most unique social and economic experiment, because its isolation, combined with its bounty of natural resources, led to labor shortages, which gave the convicts bargaining power. Fundamentally the colony *needed* its convicts, so the system was obliged to rehabilitate people. Because Tasmania took on more convicts over a longer period than anywhere else in Australia, it spent a great deal of money on social welfare, charity work, and policing.

From the beginning convicts had many rights and were protected by the law. They won significant legal cases protesting unfair treatment, and they enjoyed privileges and bargaining power that no prisoner and few free members of the working class back in England had. It wasn't hard to manipulate the system either. Convicts could work more slowly for a bad master, which would likely result in reassignment to a better situation. Some convicts even had their wives and children sent from England.

In Van Diemen's Land, 10 percent of convicts, those who had committed a second or third serious crime, were sent to Port Arthur or an equivalent prison for women, know as the Female Factory. In fact, most convicts were not institutionalized; rather, the entire island was an open prison whose inmates worked as servants in homes throughout the community or in the colonial administration. Many became educated during or after their sentence, and once released, they became teachers, surgeons, and lawyers and rose to positions of power in the government. They ran newspapers, and they even organized unions while still laboring under sentence. Once they were free, the government granted them land, animals, and seeds—enough support, pointed out criminologist John Braithwaite, that by the end of the first four decades of transportation, ex-convicts owned three quarters of the colony's land and were in possession of half its wealth. Over this time the originally mostly male population became much more demographically balanced.

Indeed, the Australian penal colony was one of the most successful examples of rehabilitation and the raising up of people in history. In 1836, when Charles Darwin's ship, the *Beagle*, stopped at Sydney, Darwin declared that it was a place of reform and reintegration, "a new and splendid country" that had "succeeded to a degree unparalleled in history." Another critic wrote that forgers became useful, swindlers taught children by their newly virtuous example, and "the thief . . . attains the magistracy."

But despite the enthusiasm from some quarters, the grand social experiment was not overwhelmingly acclaimed at the time, or since, because hand in hand with emerging personal economic power came the aspiration to respectability and the desire to cast off the criminal past. In her book, *Tasmania's Convicts*, Alexander describes how appearing clean, moral, and chaste was the most important goal of Tasmanian society from the 1820s onward. The passion for proper appearance, along with the British idea that criminality was a trait of defective families, combined to create the so-called convict stain: a stigma that marked not only the convicts who were transported but also their children, who were doomed to be inferior, immoral creatures.

That stigma didn't come from within Tasmania itself, explains Alexander, but from the British perception that a convict colony had to be an absolutely terrible place. Of course, the British didn't suddenly decide in the 1820s that the colony was stained. Even before convicts first arrived in Australia, they lampooned the idea of a criminal nation. It was only for a few brief years that the settlers lived among the gum trees free from contempt. Once it caught up with them, there was a bottomless supply.

The penal colony in New South Wales was a "sink of wickedness," crowed the British; an Australian parliamentary House of Commons would be a "den of thieves." The colonies were a cesspool, and their people were abhorrent, poisoned, blighted. When a naval surgeon told people in England that he had come from New South Wales, they would edge away from him, he said, and "check their pockets." Sometimes the people who spoke with the greatest contempt for the colony were the ones who wanted to stop transportation. They emphasized its humiliations and disgrace in order to reform the system. Although some commenters made the case that transportation had been a success, they didn't have much sway. The famous nineteenth-century social critic Alexis de Tocqueville said that essentially Australia lacked morals, and convict transportation created a society of dangerous people.

Even if the English were disgusted by the convicts, most of whom were, of course, the English working class, it was an avid, tabloid kind of disgust. British newspapers joked about convict life. Readers were titillated by reports of bestiality and cannibalism, stories that were without foundation.

British travelers experienced a pleasurable shock at the abundant evidence of Australian degradation. In the 1860s two Britons wrote a series of letters home in which they reflected obsessively on the convict stain and the way it affected everyone they met. Of middle-class Australians they wrote, "They misquoted Latin, they thought Richard II was a son of Richard I—it is scarcely to be credited."

America had thrown off English scorn when it rebelled against the mother country. But by maintaining their cultural, economic, and legislative ties to England, Australians were profoundly affected by its contempt. "In hindsight," Alexander writes, "it might have been better to brazen it out, to say that Tasmania was a fine place despite its convicts; but it would have been hard to be brazen in the face of British condescension and ridicule." By unspoken agreement everyone in Tasmania decided to simply not mention that they were all convicts, and somehow, for at least one hundred years, the strategy worked.

"Everyone had a vested interest in forgetting," Alexander told me. You can find no evidence that anyone spoke about it, she said. If ex-convicts bumped into each other and knew where the other had come from, she suspects they just winked and went on their way.

It is surely true that we are all born into a world where our parents keep something from us. It may not be scandalous or shameful, but there is a necessary filtering of information. Yet in Tasmania the silence must have been deafening. On the small island nearly all the parents were keeping the same secret from all the children.

Another historian told me that the subject of the convict past was so taboo that "not even a wife would shout 'convict' at her husband in a fight in their home." Instead everyone invented a past about being a free settler. Luckily, there were actually few barriers to appearing normal: Only the convicts at Port Arthur wore the nineteenth-century colonial equivalent of striped pajamas. Because most convicts served their time as servants in people's homes, their clothing was the same as everyone else's, if less grand. Most convicts' racial features were the same as their masters' too. Even today there is little physical evidence of the great founding population of criminals, because there were so few actual prisons at the time.

Being able to pass as a free man or woman depended a lot on one's appearance. A theory emerged among visitors in the early days that you could identify convicts because they invariably looked grizzled and beaten down, with shifty, sunken eyes and low foreheads. It was typical Victorian bigotry, and yet the theory suited many ex-convicts well: Those who did not wear their troubles or their malnutrition too obviously on their faces found it easier to pretend they weren't convicts. Ironically, because ex-convicts had access to much better food and were paid higher wages than England's own working class, it's likely that they looked much healthier than the people they had left behind in their home country.

Of course, the more traumatized people were, the more likely they would be to have "the convict look." For that small group—hard men who had seen brutal days—there was no hope of pretense. They drank and stole and set up small communities out in the bush, and by looking so unmistakably like convicts, they helped everyone else look more up-standing.

The shame and the silence wasn't exclusive to Tasmania. In Western Australia mentioning the subject of convicts was utterly taboo. In New South Wales, there were enough well-to-do ex-convicts that to some extent they were able to defend themselves publicly and fight back against the shame that was expected of them. But because Tasmania was smaller and much more isolated, the safest strategy was not affirmation but amnesia.

The only people in Tasmania who were known as convicts belonged to five particularly successful families, explained Alexander. It was their success, ironically, that kept them from escaping their past. Thomas Burbury, who was sent to Van Diemen's Land for machine breaking in 1832, went on to become a chief district constable and successful pastoralist. Everyone knew he was a convict, but given his status, it didn't matter. His family is prominent in Tasmania today.

Alexander told me about a long series of Tasmanian social maneuvers, all designed to section off the present from the past. Most obviously the island changed its name from Van Diemen's Land, well known back in England (from different sources including songs about convict hardship) to Tasmania. If visitors called the place Van Diemen's Land, they would be corrected. The word "convict" became unutterable too—and the existence of the convict colony unacknowledged. Newspapers and speakers used all sorts of linguistic dodges to avoid the C word. People wrote of "doubtful origins," "different regimes," and a "gloomy phase" in the island's history. (In Sydney ex-convicts were known as "government men.") The point, of course, was that the locals would know what was meant, while people outside the colony would not.

For a few decades after the convict system ended, there were no histories written of the colony. Even private diarists and memoirists—who were surrounded by convicts everywhere they turned and were likely descended from convicts, if not ex-convicts themselves—hardly mentioned them. No one spoke of anyone else's ancestry, let alone their own, and it was considered rude to ask. The obituaries of the best-known and longest-lived convicts did not mention how they had gotten to Tasmania, as if they had always been there; in some cases they simply lied, claiming they had emigrated.

How did people talk about Tasmania's history after transportation ended in 1856 and the Port Arthur settlement closed in 1877? In the climate of silence, completely outlandish ideas about the founding population became somehow credible. Alexander cites John West, a Tasmanian government minister and historian, who in 1852 claimed there was almost no trace left of the original convict population. They had mostly died, he said, and anyway they had borne few children. "They melt from the earth," he wrote, "and pass away like a mournful dream."

But even though personal information about convict ancestry wasn't passed down, the threat of ridicule was. In 1942, after more than twenty thousand Australians had died fighting for the Allies, Australia's prime minister said it was time for his country's troops to come home. In response, Winston Churchill, who wanted them to remain on the battlefront, declared that the Australians were "bad stock."

In the 1960s, when Joe Mauch fled the oppressive silence of postwar Germany and came to Australia, he was fascinated by the nation's convict past and asked his new Australians friends about it. But in an echo of what he went through growing up after the war, no one would give him a straight answer until someone told him, "We don't talk about that," and he didn't ask again.

Eventually the prolonged period of *not knowing* came to an end. At some point after the 1960s, when the civil rights and antiwar movements had penetrated most corners of the world, when many of the children of the last convicts shipped to Australia had died, when it became acceptable to belong to the working class, and when Australians stopped caring quite so much about what the British thought of them, the shame of having a convict relative lifted, to the point of even becoming part of the high school curriculum. In this too Alexander demonstrates that family historians led the way. Little of the history would be known if many individuals hadn't begun to research their own ancestors. Their experience shows that tracking personal history can force an awareness of the bubbles of knowledge in which one exists and the way information moves through time and space.

Perhaps the most compelling measure that some kind of great family historical cycle has been completed, or has at least moved into a different phase, is that we live in a time of trying to uncover what our antecedents so carefully concealed. A similar process took place in Ireland in 1995, 150 years after the Irish famine. Although the famine has been described as a great chasm that divided the Ireland that came before it from the one that came after, and despite the fact that its impact has been likened to that of Cromwell's invasion a century earlier, it was the subject of relatively little research and only a small trickle of books and articles. In 1985 interest in the topic began to slowly build, peaking a decade later. No doubt the arrival of a significant anniversary accounts in part for the trend, but what can explain the lack of interest before it? The famine killed one million people, more than two million people fled the country, tens of thousands were evicted from their homes, an entire social class disappeared, and poverty, fever, and mental and physical violence spread like the blight. Family members turned on one another to survive, isolating those who were ill or acting as landlords' agents. People begged for the money just to bury their dead.

A huge number of oral histories and private diaries from the period have survived. They describe experiences like coming upon a group of corpses on the road, mothers and children clinging together; a young girl standing dead against a gate; a pile of bones in a deserted cottage that suddenly moaned and moved and revealed itself to be a whole starving family. Records from government documentation, social analyses, and workhouse records exist too. Yet until non-Irish historians began to reexamine the era in the 1980s, hardly anyone spoke of it.

The Irish famine and the Holocaust were traumas of unprecedented scale, and while many aspects of the respective events obviously differ there are important comparisons. Although many second-generation Holocaust survivors suffered because their parents would never speak about what they had experienced, Jewish society has conscientiously developed a culture of remembering, lest so terrible a tragedy ever happen again. Modern historians have suggested that the newly opened dialogue about the famine may mean the Irish are finally coming to terms with their loss.

Anyone who still wonders if Francis Galton was right and criminality is heritable need only examine Tasmania to find the most complete, natural experiment of crime and genetics in the history of the world. Were the people of Tasmania a little like Robert Bakewell's Dishley sheep? Did their intermarriage over the course of a hundred years make the criminal stain darker over generations?

Apart from one rather peculiar outcome, Alexander told me, there is little to distinguish modern Tasmanian society from others that were not founded almost entirely by criminals. The settlement certainly didn't develop into a community of organized crime, where generation after generation was indoctrinated into the family business. In fact, the only tangible effect was that, for a while, the people of Tasmania were the most law-abiding in the world. According to Braithwaite, by the end of the nineteenth century Tasmania was "one of the most serene places on earth."

Despite the fact that the late-nineteenth-century rate of imprisonment was higher in Tasmania than in any nation in the modern world, the rate was lower than that of any developed country by the second decade of the twentieth century. Between 1875 and 1884 there were twenty-two convictions for murder, but over the next three decades there were none; no one was convicted of homicide again in Tasmania until 1916. No wonder everyone was willing to believe that the convicts had actually vanished. The pattern was true of Australia overall. Through the nineteenth century and into the twentieth, as the U.S. imprisonment rate dramatically increased, the number of people sent to jail decreased in Australia, and in the 1930s crime rates across the country were at an all-time low. Presumably Tasmanians were taking great pains to demonstrate that they were personally unlikely to have descended from criminals. Of course, they were all lying, but that was not a punishable offense. After a while Tasmanians relaxed a little and stopped being quite such model citizens.

Remarkably, some people still maintain the criminal stereotype of Australians, only to find that the label no longer sticks. In 2007 Australia's then most senior Muslim cleric, Taj el-Din al-Hilali, told a local Egyptian TV show, "The Anglo-Saxons arrived in Australia in shackles. We came as free people. We bought our own tickets. We are entitled to Australia more than they are." Al-Hilali was well known for making inflammatory remarks, but the reaction to this particular statement was not outrage but hilarity. He later insisted that all he meant was that "we love Australia so much that we choose to be here. We were not forced to be here."

I finally learned what Michael Deegan looked like when a second cousin showed me a photo of him. I hadn't even imagined that one might exist, and it was an amazing feeling to finally see the face of the man who had survived so much. At Point Puer he learned how to be a sawyer. I found out much later that somewhere along the way he lost a hand. He was released in 1848 and moved to the southernmost mainland state, Victoria. He looked small, even impish. He grinned at the camera, and it was easy to detect a twinkle in his eye. I also saw a photo of Ann McGrath, convict wife, possible convict herself, mother of ten. She had light eyes, like her husband, but their demeanors couldn't have been more different: She looked exhausted.

It looks like Deegan's strategy succeeded. It took a few generations, but the pain is gone, and the history that was actively, perhaps desperately, hidden has faded into a history that has simply been forgotten. My own nonscientific observation is that in this respect my family doesn't seem to be very different from the families of people with whom I grew up, some who immigrated a long time ago, some who arrived more recently, and presumably only some of whom have a colonial convict in the family tree. The majority of Deegan and McGrath's great- and great-great-grandchildren are law-abiding and employed. Two are professional comedians. One is a magistrate.

It's fashionable now to assert that genealogy reveals much more about the genealogist than about his or her ancestors. No doubt there is a kind of Rorschach process involved, whereby the crowd of names and facts that one is able to pull together from whatever records remain looks as if it had a certain shape. Some people in my family tree stand out to me because what I learn about their lives feels as if it has relevance to my own—I found out about Julia and all her children when my children were a similar age. But this isn't therapy; it's the history of human lives. Michel Deegan was real, a man who loved and suffered, and the events of his dark and extraordinary life shaped him far more than my discovery of them has shaped me. I am, of course, curious about his legacy to all his descendants, myself included—not just the poverty and the crime but the secret he guarded as well. His secrets, and those of the other inhabitants of Van Diemen's Land, shaped its culture for a long time, but what did the actual experience of concealment feel like within families?

If Michael Deegan and Ann McGrath's children were not aware that their parents had been convicts, it would be consistent with what the historians say—most people simply didn't speak of it. My father was raised in part by his grandmother, Michael Deegan's youngest daughter, and he remembers her as a particularly refined and ladylike woman. She trained as a teacher and then raised five children. She was also a twin, and her sister became a nun. Did she know?

I contacted some of my father's cousins, who were as surprised to hear the convict story as he had been. But they agreed that the convict's daughter was renowned for her class and elegance. Did her parents teach her to be that way? Did she learn to be like that to distinguish herself from her parents? Was she a naturally graceful woman who was also lucky enough to grow up in a country where there was plenty of food and education for most people?

I was a little nervous to tell my father about Michael Deegan, but the changing times had changed him too. He was completely surprised. He had never heard the slightest whisper of the truth, but he had no trouble accepting it. He was sad for Deegan and McGrath, and he was full of compassion for their situation. He didn't think that anyone in his parents' generation knew about our convict.

I found out where Deegan and McGrath were buried, and one day my parents and I drove to the small country cemetery outside of Castle-maine. On a quiet road at the base of a steep green hill, they shared a grave with their son, his wife, their youngest daughter, and, as the grave-stone said, "others." Their stone was the smallest marker in the grave-yard, but a beautiful tree leaned over where they lay, shielding it from the elements.

It used to be that people could expect their illicit pasts would die with them, but the personal computer and Internet revolution have changed all that. Digging up records was once laborious, specific, phys-ical work, but now it often just involves opening one's laptop. The easy availability of all records means it will be harder and harder to invent your own past. For good and bad, the new historical transparency brings new responsibilities with it too.

Chapter 6

Information

The greatest crisis facing us is not Russia, not the Atom bomb, not corruption in government, no encroaching hunger, not the morals of the young. It is a crisis in the *organization* and *accessibility* of human knowledge. We own an enormous "encyclopedia"—which isn't even arranged alphabetically. Our "file cards" are spilled on the floor, nor were they ever in order. The answers we want may be buried somewhere in the heap, but it might take a lifetime to locate two already known facts, place them side by side and derive a third fact, the one we urgently need.

Call it the Crisis of the Librarian.

We need a new "specialist" who is not a specialist, but a synthesist. We need a new science to be the perfect secretary to all other sciences.

But we are not likely to get either one in a hurry and we have a powerful lot of grief before us in the meantime.

—Robert Heinlein, "Where To?" (1950)

There is an extraordinary group based in the United States that sends its representatives all over the world to gather information about the life histories of human populations. They make deals with states and churches to view the records of their citizens, both living and dead. Once inside the libraries and the vaults and the back rooms of the world's archives, they set up their cameras and painstakingly photograph every birth certificate, marriage contract, and death certificate they can find. The images are sent back to their headquarters and stored in a vault that was hollowed out of a granite mountain in the Utah Rockies by their forebears, whose names are also stored there. In their theology family is sacrosanct, and scholarship and genealogy are essentially holy activities.

In October 2012 I drove through a steep-sided pass to see if I could find the Granite Mountain Records Vault. The scrubby plants by the road were mustard yellow and rust, but the autumn breeze that blew through the car window still had a touch of warmth. On the south side of the pass, the rock's geological layers pointed eighty degrees up at the sky. A few miles in, a rattlesnake lay coiled in the middle of the road. Thanks to old photographs and current Google satellite images, I found what I was looking for and got out to gaze up at the concrete arches that lead into the vault, built by the Church of Jesus Christ of Latter-day Saints (known as the LDS) in the 1950s.

In the 1970s journalist Alex Shoumatoff described the vault's six 200-foot-long chambers and three 350-foot-long corridors; its air-filtering, ion-detection, and smoke-detection systems; its meter for tracking movement in the earth's crust; and the spring-loaded blast locks that will seal it in the event of a nuclear explosion, the natural temperature and humidity levels of the chambers preserving the film and paper records that remain. At the time of his visit the most striking wonder of the vault was the number of drawers that filled its chambers and the massive number of analog documents they held. Since then, access has been restricted, and the media and the public have been shut out. Still, the vault is one of the informational wonders of the world, and I was determined to see it, if only from the outside. According to the Mormons, millions of personal records are stored behind a fourteen-ton steel door, which stands three hundred feet inside the mountain.

A week earlier in Colorado I drove into the same mountain range hundreds of miles east and placed my hand in the imprint left by a three-toed dinosaur one hundred million years ago. On Dinosaur Ridge, where the stegosaurus, allosaurus, and apatosaurus were discovered, clearly defined rows of footprints stomp up the rock wall, a busy jumble of moving feet from the shoreline of an ancient sea. The entirety of human history may be more ephemeral than one of those steps, but in the meantime the architects of the vault are leveraging the mountain against that possibility. Their goal is to preserve not just human history but all the individuals in it—at least, those whose existence has been noted on paper. The LDS is not the only American institution caching data in the mountains, either: Rumor has it that a U.S. government facility and various private facilities are dug in along the same road. Still, the Mormons think as hard as, probably harder than, anyone else in the world about what it means to keep facts alive, or at least to keep them accessible to the living, and the phenomenon they have built out of granite, microfilm, machines, and software is as mind-bogglingly ambitious for our century as the flying buttresses and gargoyles of Notre Dame were in the twelfth century.

Even as a large branch of American genealogy sheared off at the turn of the twentieth century into a mad eugenic scheme to reshape the human race, the Mormons got on with their mission to gather and share records. Around that time Mormons whose ancestors had come from Europe could find out about their forebears only by traveling back to their home countries and transcribing whatever information they could find. As a way to assist its members, the church began to send representatives to locate collections of records, copy them *all*, and bring them back to Utah. In the 1920s the church began recording the genealogical information it had gathered on index cards, and in 1938 it started to make copies on microfilm. Eventually the microfilm was circulated to thousands of Mormon libraries throughout the world. By the 1950s the church elders faced an ever-growing pile of film, and in the wake of the great destruction of records in Germany in World War II, they started to dig into the mountain to store it safely for posterity.

The mountain now holds parish records and old English manuscripts dating from the 1500s, including records from London, when civil registration began in 1837, and copies of *jai pu*, Chinese family records, which date back before AD 1. Overall the data the Mormons have gathered is equivalent to thirty-two times the amount of information contained in the Library of Congress—and the church adds a new Library of Congress's worth of new data every year.

This massive infoverse exists to serve Joseph Smith's late-nineteenth-century teaching that church members should offer baptism to dead relatives. Because members may only carry out the rite for their own ancestors, all church members now spend a great deal of time tracing their lineages back through time. Have humans ever built anything of this magnitude without an eye on the afterlife?

Fifteen miles away from the vault, in the clean streets of Salt Lake City, I met with Jay Verkler at the Joseph Smith Memorial Building. Built originally as a grand hotel in 1909, the structure stands next to the white, Disney castle–like Mormon Temple. When we met, Verkler was the CEO of Family Search, the Mormon organization that manages the vault's records and promotes genealogy throughout the world. Once a gifted twelve-year-old who wrote software for the bank where his father worked, Verkler became a Silicon Valley entrepreneur until the church's elders summoned him back to Salt Lake City. Verkler is of an imposing height, and he has a thick helmet of blond hair (which, at a recent genetic genealogy conference hosted by the LDS, had its own Twitter feed, @Jay-VerklersHair). He looks exactly like the kind of modestly presented, clean-living Mormon missionary you might find knocking on your front door. His command of the intricacies of information storage in an ever-decaying world combined with an implacable commitment to the eternal ideals of the church make him a powerful presence. More than any other organization his church has shaped how genealogy is practiced in the world today.

"The core concept of why this church cares so much about genealogy stems back to the notion that families can be eternal organizations past death," Verkler explained. "Members of the church seek out their ancestors because we think we have a duty to them to help them understand this gospel that we understand, and we think we can actually be together."

The idea was magically appealing. At the time, my own boys were so young that I could scarcely imagine a time or place where I would not be present for them. As Verkler continued to talk theology, I mused at how brilliant a basis this was for a religion. What parents would not want to believe that they could be with their children forever?

Of course, if entire families are destined to be together in the afterlife, that would include parents and siblings and their spouses and children, aunts and uncles, and in-laws. Is this afterlife going to look like some kind of celestial neighborhood where the streets map out bloodlines, with entire apartment blocks assigned to close families? Or will it be more like a perpetual Thanksgiving feast designed by M. C. Escher after a bad night's sleep?

"We're not quite sure how it's going to work," Verkler admitted. "It's not going to be like one big group family, but we think those connections will still exist in the afterlife."

The LDS philosophy is about not just the next stage of existence but life before the afterlife too. "We think there's a strengthening of you as a human when you know who you came from and where your roots are and when you respect that part," said Verkler. He surely speaks the truth, because some of the Mormons I met in Salt Lake City were the friendliest people I have ever come across, respectful and polite to a most disarming degree.

Over the last ten years Marshall Duke, a psychologist from Emory University, has explored the value of family history in the lives of children. He developed a list of twenty questions such as "Do you know where your parents met?" "Do you know which person in your family you most look like?" and "Do you know some of the jobs that your parents had when they were young?" Duke found that the higher children scored on the family-history test, the higher they also scored on measures of self-esteem and self-control and the lower they scored on anxiety, among other measures. Duke even looked at children who experienced the terrorist attacks in the United States on September 11, 2001. Even in this extreme case, knowledge of family history appeared to indicate how resilient the children were in the months that followed. Duke explains that it's not necessarily the facts of the family that give children these qualities but the fact that, if children can answer these questions, it usually means that they have strong connections with mothers and grandmothers and that significant amounts of time have been spent communicating at family dinners and on family vacations. All the stories of a family add up to what Duke calls an *intergenerational self*, which he associates with personal strength.

All the industry that the Mormons have devoted to assembling genealogical records is not just for church members. "We provide our records for everybody," Verkler explained. "We think that it's doing good for the world." Accordingly, there are more than 3,400 Family History Centers in the world. They are a sacred municipal library system, and anyone who wishes to research his family history can make use of them. Smart, kindly people will help him search historical documentation such as birth records, death certificates, land records, and any other document that might establish a genealogical connection. A borrowing system between the centers and the main Family History Library in Salt Lake City means that if your local center doesn't have the record you are after, another might be able to copy it onto a disk and send it.

In this respect too the LDS differs from all other religions. Its kind of twenty-first-century munificence requires an extremely sophisticated understanding of informatics and digitizing. Trying to determine and then store everyone's name and existence for perpetuity is also an insanely costly process. Today the Church has 220 data-gathering teams in forty-five countries that are making digital copies of new records. They are also converting 2.4 million microfilm records into a digital format. The LDS drove microfilm technology in the twentieth century, and today it is a leader in digital data storage. Its digital camera operators photograph records and get those images online within two days, and then an enormous army—that is to say, hundreds of thousands—of volunteers index the files and make them searchable. The Mormons were crowdsourcing long before the word was invented.

The last time I visited the church, it was deeply engaged in its biggest project to date—a joint effort with the national archives of Italy in which more than one hundred Italian state archives gave the LDS teams access to all the birth, death, and marriage records from 1800 through to 1940. LDS photographers have taken more than three million images of the files, which recorded the lives of five hundred to seven hundred million Italians from the nineteenth and early twentieth centuries. They included people who lived before the invention of photography, people who watched their children die of the flu in 1918, and people who years later themselves died at the end of World War II. It is the most definitive collection of Italian civil records in the world.

The church's most ambitious project is its online tree. Anyone who logs in to Family Search may record and research his or her family history there, but what distinguishes this tree from all the other online services is that the church is trying to connect all the branches, using its massive records and the activities of users to build a big tree of all of humanity. The endeavor must be, to some extent, possible. If anyone has the records to create this structure—a family history of all of the documented individual members of the human race, this group does. But the distinctive element of the LDS tree is that it's collaborative: People can log on and add names and link them to documents and write personal stories—and once they have done that, their fifth cousin once removed may also jump online and edit that information, changing a relative's name, linking it to other documents, or deleting the story altogether. No one I spoke to at Family Search seemed to think this would be a problem, but surely everyone's version of her own family is different from that of her cousins?

Still, even if the online tree is in constant flux, the names and lives of millions of people will stay safe in the vault long after the names chiseled into all the world's gravestones have eroded to nothing. The Mormon records will last for a very long time, at least until a natural disaster occurs, or maybe until some point in the process when a human being makes a mistake.

In 2004 Cyclone Heta, a record-breaking category 5 storm, hit the South Pacific island of Niue, the world's smallest nation. Heta's wind gusted at up to 177 miles per hour, and the waves it sent over the island have since been described in a technical report as "extremely very high." The storm washed seventy homes and businesses over a ninety-foot cliff into the sea, and salt carried by the wind destroyed crops and vegetation. Overall eighty million dollars' worth of damage was done, and not just to the residential and commercial parts of the island. All of the small nation's birth, death, and court archives were destroyed. But because the Niue government had stored many of the island's genealogical records with the LDS in the Granite Mountain Records Vault in 1994, it was able to retrieve copies of them.

The biggest problem with data, once it is collected, is how to preserve it. Because we mere mortals have only the most tenuous mental grasp on the passage of time and our tiny place in it, we tend not to recognize the basic existential truth that, as time passes, stuff gets lost. People forget where they put things. Nations forget where they stored things. Important documents are thrown out. Other important documents are suppressed. Buildings are bombed. They flood or burn down. The 1906 San Francisco earthquake destroyed most of the city's birth, death, and marriage documents. In 1922 the Public Records Office of Ireland was incinerated, and now only a few documents from before that time remain. Sometimes we lose not information itself but the information that would have helped provide a context for the information that we do possess.

Preservation is made even harder by how rapidly technology rolls over. It is often the case that shortly after we invent a new process for recording information, a better one appears, and all the data so carefully stored in the original medium must now be transferred. Consider the fact that as a child, the typical forty-five-year-old used to listen to music on eight-track tapes or vinyl albums. He then recorded his albums onto cassette tapes. Later he tossed out both the albums and the tapes and bought the same music on compact discs. Now CDs are rapidly disappearing, and the same people have begun downloading their music as a digital file. If someone still happened to have an eight-track copy of, say, the Beatles' White Album, it's unlikely he would still have the machine to play it on. If he ever wanted to listen to his old recording of "Blackbird" again, he'd have to hunt down an eight-track player or build one himself.

The problem of rapidly evolving technologies or "digital migration" was rather alarmingly illustrated in England in the 1980s with a considerably larger amount of information. Actually, it began in 1086 with the Domesday Book. The first public record ever made in England, the Domesday Book was instigated by William the Conqueror, who wished to take a census of his people and, more specifically, their possessions. He dispatched men to all corners of the realm to record how much land and how many animals were held by over thirteen thousand of his subjects. The goal, of course, was to tax them. Once all the king's nobles, church officials, and common landowners were surveyed, their information was recorded in Latin on a sheepskin manuscript in two volumes that came to be known as the Domesday Book, or Book of Judgment. Surveyors were given considerable power, and once they recorded someone's holdings, their assessment was the final word—forever. Many of the places recorded in the Domesday Book still exist in England today, even if their names have changed, and a number of families can track their lineage to individuals cited in it. The book was even used in the 1960s in a court case over ancient land rights.

Almost a millennium after the Domesday Book was compiled, the British Broadcasting Corporation and a few computer companies got together and decided to make a second installment of the first great census. The goal was to capture all aspects of life in the United Kingdom at the millennial interval. Between 1984 and 1986 over one million people contributed to the project by filling out a survey. Photographs and video footage were collected, and virtual-reality tours of streetscapes were created. Schoolchildren from all over the country wrote entries about where they lived. One child in Orkney, a small group of islands off the coast of Scotland, reported on a hurricane that hit the area in 1952: "Henhouses with dazed or dead hens were blown out to sea. Some hens had all their feathers plucked by the wind. Another crashed through a farmhouse window at a terrific rate, landing in a box-bed. The occupants were astonished."

The project was so ambitious for its day that the researchers who initiated it had to invent new technology to contain all the material that had been gathered. Eventually the second Domesday collection was stored on laser discs, an optical storage medium that was a predecessor of CDs and DVDs. To read the discs you needed an Acorn BBC Master computer, supplemented by a few other pieces of machinery. If you wanted to navigate through the material, you also required a Master keyboard and a specialized trackerball.

Like most other technology from the 1980s, laser discs have been supplanted many times over. So, even though it was thought at the time of its creation that the shiny new laser Domesday Book would outlast its sheepskin antecedent by many thousand of years, the discs barely lasted fifteen. By 2001 no one even knew where to find an Acorn BBC Master computer. The company that made one of the additional pieces, the LV-ROM drive, had only manufactured one thousand of them to begin with, and the laser discs themselves became unstable. In fact, most of the technology needed to access the second Domesday Book, including ye olde trackerball, became so obsolete so quickly that less than twenty years later no one knew if any of it still existed.

After much angst, part of the second Domesday Book was retrieved and put online in 2004. When a key team member died, the project ground to a halt. In 2011 a different team published some of the 1980s book online, and for six months the BBC invited people to submit twenty-first century updates to the information gathered in the 1980s. Apparently access to much of the original content is still restricted because of copyright issues.

At around the same time, a similar problem arose in Iceland. As part of an effort to digitize all its censuses, the government had to retrieve the punch cards that had been used to record the data in the 1960s. But they were impossible to read, not because the cards themselves had degraded but because no one in Iceland still had a punch card reader.

Even when people seriously consider the preservation problem in a project's design stage, it's not easy to avert. In the early twenty-first century the International Atomic Energy Agency (IAEA) assembled a team to investigate how people of the present might best communicate key information with people of the future. How should we mark radioactive waste so that generations in years to come do not accidentally stumble across it? Initially the team wanted to create a physical record that would last as long as the radioactivity, potentially for tens of thousands of years. But teams from Sweden, Canada, and Japan had already tackled the problem, and the experience of the Japanese team suggested that this approach would not be fruitful: The Japanese had created a silicon carbide tile that measured about twelve square centimeters and looked much like the kind of tile that might be found on a bathroom wall yet was incredibly hard and would not erode, which meant it could be buried in the ground. The team etched the necessary warnings on it with a laser, so the writing would never fade. There was just one problem: If you dropped the tile, it shattered.

Actually, there were two problems. Archivist Gavan McCarthy, who worked as a consultant on the (IAEA) project, explained: "If *well housed*, the tiles would last ten thousand years. Which was not bad. But then that just raised the whole question of: In ten thousand years if somebody discovers this, could anybody actually understand what was on it?"

"What a community needs is continuous knowledge of the existence of the material," McCarthy said. "If it's accidentally dug up or just comes to the surface through some volcanic event in the future, that would probably melt the tile anyway."

Counterintuitive as it may be, it seems as if the record-making method we invented a few thousand years ago—that is, writing the words of a common language with a handheld marker—is still the most durable. Actually, this technology dates even further back than that: Paper degrades quickly, animal-skin manuscripts last longer, but the world's oldest records were carved into or painted on rock. One of the oldest records found in many places all over the world is called a cupule, a round indentation carved into rock. No one knows what a cupule actually *means*, but because they exist we know at least that the people who created them once existed too. The oldest known marks were crosshatchings made in rock and left in a cave in South Africa seventy thousand years ago. (There is, in fact, one better way of preserving information, but we didn't invent it—see the epilogue.)

In the end, the silicon carbide tiles did not go into production. "Without continuous knowledge, then all systems of knowledge are fatally flawed," McCarthy said. "The reality is that all you can do is hand on as much knowledge as you can to the oncoming generations to give them the best chance you possibly can to do what they can."

McCarthy's rule applies to all culture. Imagine if Shakespeare had composed his work on laser discs. What if the Bible had first been recorded onto eight-track tapes? Preservation isn't just about the durability of records; it's about the durability of the people who care about the records. At a certain point after Shakespeare's plays and the books of the Bible were created, they became so popular that no central body was required to plan their migration from one technology to another—it just happened. Whether for pleasure, out of righteousness, or for profit, generation after generation has engaged with the texts and transferred them from whatever medium they found them in to the one they preferred. From his original draft on parchment, Shakespeare's plays have over the centuries been rendered in many formats. A copy of his complete works came to me free on the iBooks app of my iPad.

While the LDS is transferring its data from microfilm to digital storage, it is not making any assumptions that digital will be the final version. "People have been talking about digital preservation for a long time, but no one has actually been building the systems to do it," Verkler observed. "We think that polyester-based microfilm will last for somewhere between three to five hundred years. For digital, the bytes will rot off the media that you create—within ten to twenty years on DVDs, for instance. All these CDs that people are burning and think that they're going to last for a long time, they're not. They're going to be unreadable."

What if there were a huge natural disaster, and everything *outside* the Granite Mountain Records Vault were destroyed? Future historians could retrieve the mountain's records and re-create many hundreds of years of demographic history. Would they also discover that most humans from all of history were, in fact, Mormons?

In the 1990s a Mormon group started working its way through all the names of the victims of the Holocaust, apparently baptizing them into the LDS. The controversy that erupted was resolved by a 1995 agreement between Jewish leaders and the LDS, whereby the church agreed to remove the names of posthumously baptized Jewish people from its records. But in the years that followed many Jewish names found their way back into them.

In 2003 an Armenian group protested that the LDS had baptized by proxy notable members of its community as well. In 2008 the Vatican sent a letter to parishes all over the world asking them to not share their records with Mormon genealogists. In 2012 it was widely reported that Anne Frank had been posthumously baptized into the Mormon Church. Similar stories emerged. Stanley Ann Dunham, the late mother of Barack Obama; Daniel Pearl, the *Wall Street Journal* reporter who was abducted and murdered in Pakistan in 2002; Adolf Hitler; Simon Wiesenthal, the Nazi hunter; and Steve Irwin, the Australian TV naturalist, had all been baptized.

I asked Jay Verkler about proxy baptism. It was a misnomer, he explained: Members of the church *offer* baptism to their ancestors. These ancestors are then checked off a list that notes that they have received an offer. That list is different, he said, from the "Members of Record" database, which includes only the names of people who have officially, during life, accepted such an offer.

Nevertheless, Verkler said, Frank had probably been offered what the church calls proxy ordinance about one hundred times. Members are supposed to offer proxy ordinance only to their own ancestors, but the policy has occasionally been abused. "What happens is that a member is reading about Anne Frank and [he] says, 'Boy, I hope someone has made this offer to her. I think I will.' And they go and they take care of it. Sometimes people get a little misdirected there."

Mormons, explained Verkler, have warm associations with the idea of baptism. He understands that many Jews do not. "There were some really awful things that have been done to the Jewish community. Jews were forced to be baptized or burned at the stake, so 'baptism' is not a happy word. We didn't understand that for a while, I think, culturally." (As one Jewish genealogist confirmed to me, "The whole idea of proxy baptism is incredibly offensive for Jewish people.")

"On the other hand," Verkler said, "if you think about other religions that light a candle and say a prayer for someone, or create a prayer for someone who is deceased, it's not a unique pattern, so that same kind of motivation is what I think motivates people."

The same motivation may be involved, but as many Jews have pointed out, when they light a candle, they don't make a record of it. The practice remains a point of tension between the two faiths, especially as there is a large Jewish genealogical community that relies on the resources created by the LDS.

Future historians of the Granite Mountain Records Vault may also be surprised to find that only heterosexual people married and had children in the early twenty-first century. Within the last two years, a growing series of online complaints have noted that people who want to record marriages of family members who are the same sex cannot because the software won't record the union. Which is to say, the family tree database won't allow users to report a marriage unless it takes place between a man and a woman. If this is the only database that survives a catastrophe, it will offer a skewed picture of life in our time.

Remember Essie Mae Washington-Williams, the illegitimate daughter of Senator Strom Thurmond? She said, "There are many stories like Sally Hemings and mine. [Hemings, a slave, had children fathered by the United States president Thomas Jefferson. See chapter 11.] The unfortunate measure is that not everyone knows about these stories that helped to make America what it is today." What America is today is a nation in which same-sex marriage has been recognized in seventeen states and eight Native American tribal jurisdictions. The federal government of the United States recognizes gay marriage, as do those of at least nineteen other countries. In the United States alone there are at least 220,000 children being raised by same-sex couples. But because the LDS software won't register these unions, all those American stories will have been lost, and the database of millions is no longer a real record, because it doesn't record what's real.

If it weren't for Ancestry.com, Geoff Meyer, who was raised in the awful orphanages of the twentieth century, would never have found the smallest scrap of information about his father. Meyer's experience may be unusual, but he is one of millions for whom the organization fills a need. Based in Provo, Utah, with a large office in San Francisco, the company's TV ads feature friendly middle-aged people stumbling on the fact that their grandparents' wedding took place just a few months before their first child's birth; or becoming overwhelmed with emotion when they discover a fact about a loved parent who is gone; or being thrilled by a coincidental crossed path, like the fact that they lived only four blocks away from an ancestor that they never met.

The Mormons help many people around the world because of their spiritual mission. The mission of Ancestry.com may be rather more secular, but it is no less powerful. As one genealogist observed to me, it's not about the "begats" anymore; it's more about the stories. Certainly Ancestry.com has addressed that need. Most people in the field now are talking about taking it beyond the scholarly pursuit. "That's the classic genealogist," Dan Jones at Ancestry.com told me. "If we're saying that the interest in who we are or where we come from, the interest in identity, is universal, then the interest in trolling through microfilm certainly isn't universal, and it's certainly not endemic throughout every life stage."

Ancestry sponsors many genealogy expos all over the world, some of which are attended by hundreds of thousands of people. Jones, who is often on the front line at these events, says the fascination with genealogy is expressed in different ways in different cultures. As closely related as the Americans and the British are, there are still powerful distinctions in what appeals to them about genealogy, or at least in what they say appeals to them. Advertising for Ancestry.com in the United States highlights the emotional or the scandalous nature of revelations. But, explained Jones (who is from Wales), "You put that in front of the British audience, and they're like, *Oh my God, you got to pay to be emotionally damaged? Why would I do that?*" Ancestry runs focus groups in all of its markets to determine how particular cultures prefer to think about their personalized past. "The British take on it," Jones told me, "seems to be, *No, it's not; it's not emotional at all. I just want my facts here on this paper, and I want to be able to give it to my offspring and my grandchildren and tell them where they came from. It's not the story of me; it's the story of them.*" He tapped the table to make the point. "*Those people on those pages.*"

At expos Jones often finds himself talking to people who are keen to connect. He helps them try to find relatives in the database, and it can end up being quite an intimate way to talk to strangers. "You're doing lookups for people and you inform some lady at three in the afternoon that it's highly likely that her grandfather was a bigamist. You can't say it's true," said Jones, but "you explain to her how difficult it was to get divorced in 1898, and yet when people got to the point where they couldn't stand the sight of each other, what did they do? Quite likely you just move twelve miles away to another town and have another family.

"People *want* to talk about it," said Jones, "because so many of them turn up. They may be fully aware their grandmother was a bigamist, but they want to speak to you about it and tell you about their experience of finding it. I think that goes to the heart of what Ancestry.com is for people. It's a way for them to find their place in the world."

Still, Ancestry.com isn't just a historical matchmaking service—it's a massive data company, because modern genealogy, of course, is the Big Data of little people. And however much eugenics and Nazi catastrophes have shaped the distaste and anxiety people feel about family history today, and however much the New World still burns to declare its independence from the old, genealogy companies have quietly and steadily expanded to become some of the biggest data organizations of the twenty-first century.

Ancestry.com began in the United States and now has big followings in Canada, Australia, and the United Kingdom. When I visited its San Francisco office, Dan Jones told me that the company was "exceptionally well represented in the Western English-speaking world." This is not merely a matter of particularly interested audiences but also reflects the fact that, because of their political structures, these countries also have good, easily available civil records. Ancestry.com also has a presence in many other countries, including Sweden (Ancestry.se), and is currently trying to expand into Mexico.

Overall the company's holdings include twelve billion records. (It defines a record as a piece of information, like a birth date or a marriage location.) Like the Mormons, it makes copies of government, census, and other civil records from all over the world. In 2012, when the 1940 census was released to the public, it took Ancestry.com less than four months to get the entire 132 million records online—not just names and dates but all the information recorded in the census. At the beginning of this chapter I quoted the famous science fiction writer Robert Heinlein, who worried that the production of information in the human world far outpaced our ability to organize and digest it. Heinlein issued his warning in the 1950s, long before personal computers and the Internet and the genome and Big Data, all of which have made the problem exponentially larger. So far Ancestry.com's solution has been to get millions of people from all over the world to use their personal knowledge and detective skills to connect and sort much of the data themselves. A considerable portion of its twelve billion bits of information are documents added by users.

As far as Ancestry.com's total holdings go, they are shaped by money and the way it flows. China, for example, has always been one of the best record-keeping nations, and although there are more than 1.3 billion Chinese people in the world, there aren't enough to create a market for Western-style family history. "If you look at the demographics," Jones said, "42 million Americans have German ancestry. There's 3.7 million with Chinese ancestry. We could acquire a ton of Chinese content, but the East Asian community in the U.S. isn't that big outside San Francisco, and if you look at the UK, nearly all the migration comes from Hong Kong."

This is all taking place in a world where there is exponentially more data than there ever has been. Facebook is another example of the way that millions of people across the world now document and share their own lives. (Indeed, many genealogists believe Facebook, where users connect with friends and post updates about their personal lives, is the beginning phase of developing an interest in family history.) Self-recorded personal data is just one side of the Big Data coin: The other is the data that everyone else is keeping on us, whether it's the retailers who track our purchases, the insurance companies who monitor our health, Google, or the government. The biggest Big Data scandal of 2012 was the revelation that what the people with tinfoil on their heads have been telling us for years is more or less true—*they* are listening to *everything*. The NSA has been surreptitiously monitoring the phone and Internet use of millions of ordinary Americans as well as of foreign powers for years.

Of course, what makes all the stories, events, and moments in our lives available as data that can be published is the fact that all this information is now digital, and none of this would be happening without the Internet. A 2011 study by researchers at the University of California reported that 2002 marked the beginning of the digital age: It was in that year that digital storage capacity became bigger than analog for the first time. Now, according to the researchers, 94 percent of all our stored information is digital.

By making copies of records and digitizing them for their users, family-history companies are in many cases the only ones making an effort to keep this data alive. In some countries it's not an easy task. In Italy, as in many European nations, the communities are so richly endowed with records of the past that they don't spend a lot of time looking after them. One Ancestry.com representative showed me a photograph of some ancient Italian civil registries piled up on the cistern of a flooded toilet. Artifacts like these are endangered all over the world.

As enormous a quantity as Ancestry.com's ten billion records seems, they are only a fraction of the records that exist. There are millions of undigitized documents in large archives all over the world, and frankly, it's hard to overstate their inaccessibility. To find something you often have to access a search aid that is in another room, if you are lucky, and potentially in another building; you will probably have to wait for the one day of the week (or even the month) that the person who is a specialist in interpreting those documents is paid to come in.

Yet someday all of these documents may actually be connected, or connectable. It wasn't until the existence of the Internet that we even conceived the notion of considering them as a single body of information. In fact they constitute an enormous infosphere that hangs quietly looming next to the world of people. As we find ways to read, organize, and connect all this data, we can map patterns, develop insights, apply analysis, and make predictions. This is true of any information we can obtain about the past, but when it is digitally stored, it can be searched almost effortlessly, uncovering data points that would have taken years to discern in physical media, and applying the tools of data analysis can produce more data about our data. This is probably especially true for family history, because a family is essentially a network where individuals are connected by bloodlines.

Kevin Schurer of the University of Leicester, previously director of the UK Data Archive, examined census data for his PhD in 1988. While demographers and historians have often used such records, because they had to be transcribed and then physically input into a computer, "it was a very time consuming process and it limited what you could and couldn't do," Schurer explained. Recently Schurer made a deal with the United Kingdom's largest genealogy company, FindMyPast, whereby he cleaned up and coded about 215 million records from their late-nineteenth- to early-twentieth-century census data and in return was able to use the data for his research and to make it available to other academic researchers.

"When you have a hundred-percent count data," Schurer said, "that allows you to start looking at things which would be too small to analyze locally. To give you an example, one of the things which are captured in the censuses is disability, whether or not you're blind, deaf, dumb, or whether you suffer from a mental frailty. This has never, ever been fully analyzed. Why? Because when you look at any one place, invariably you just might only have one or two blind people, so you can't actually look at it until you look at the whole country."

Schurer began to draw maps of the incidence of deafness, mutism, blindness, and mental frailty. The preparation of the data took four years, and the analyses have only just begun. But it's already clear that there will be rich findings. "If you think about it, you should expect this to be evenly spread across the country, but it isn't. It is actually very geographically skewed," Schurer noted. He suspects that the occurrence of blindness and deafness may be explained not by genetics but by occupation: The mining industry likely contributed to their high incidence in some locations.

The Big Data analysis of little people could change the way a subject like migration is studied. Because you can better chart in-migration and out-migration patterns, said Schurer, "it allows you to understand the link between migration and economic development much more fully." Data also enables tracking in much greater detail of how the economy changed throughout the nineteenth century, rural depopulation, and household statistics and fertility. In fact, Schurer said, "You name it, these data will give us a much greater idea of historical processes in the past for several areas of research." In 2013 another researcher used family tree databases to build an enormous family tree, including one pedigree that begins in the fifteenth century and has thirteen million people in it. The anonymized tree is available to researchers who want to study demographics, longevity, and fertility.

The typical deal that Ancestry.com makes with archives is to digitize their records while leaving the original physical records freely access-ible, so people who prefer to search by hand can still do that. But the company charges for online access. Everything Ancestry has created is added to the original record series, Jones said. Often there's a free, somewhat limited public-library online version as well. While access to such archives is critical for both the government and academia, said Jones, the overwhelming majority of the stakeholders in any archive are genealogists, so their most vital users are personal historians. With governments and corporations throughout the world pulling funding from their archives, if Ancestry.com, Geni, and other genealogy companies weren't investing in them, no one would be.

But is that a good thing or a bad thing? The past is, of course, gone, and all that we have left is information about it—which means that our relationship with the past is a relationship with *records* of it. But if that is the case, does that mean that these companies are charging us admission to history?

In fact, many archivists believe that Ancestry.com and their kin have made a positive contribution to the world of records. By streamlining and packaging documents so that millions of people can find their antecedents, the companies have created a completely new way of interacting with history. Still, archivist Cassandra Findlay told me there has been growing concern about the presence of these companies in government archives. When utilities, toll roads, or other public assets are sold off to private concerns, there is usually considerable public debate, said Findlay, yet on this issue there has been none. Over the last few years, she explained, "Companies have stealthily crept in and struck deals on a one-to-one basis, and it has been invisible to governments and to the public." Part of the problem is that it's not yet entirely clear what we might be losing if governments allow archives to be completely shaped by market forces rather than considerations about heritage. But even if the more specific risks are unclear, the general concerns should be obvious. The records, said Findlay, "represent access to memory, rights, entitlements, and accountability. They really belong to everyone. I'm not saying we shouldn't strike deals with these companies. But there should be more conscious decision making."

Still, if companies were not able to use the records to make money, few people would ever see them. It may not be too long before the value of records becomes more obvious to everyone. In just the last few years a number of very different projects have demonstrated what data-based history can tell us—even when it is data that's been pulled together by the most maligned researcher of all, the family genealogist.

Yaniv Erlich, a researcher at the Whitehead Institute, collaborated with Geni.com, the genealogical Web site, to take data from forty-three million genealogical profiles and map it. Erlich and his colleagues plotted the birth dates of all the individuals in Geni's ancestors database on a map of the world. It turns out when you put this data together, you get a dynamic and fairly accurate picture of the history of the world, or rather of historical events that have been written down.

In order to make the data visual, the researchers divided it up into ten-year segments. They then created a video of a world map, on which a tiny dot glows to represent each individual born in that decade (whose birth was recorded). Each dot eventually fades as the next "generation" appears in the subsequent decade. The pulsing glow that results charts the movement of people all over the world. At the end of the fifteenth century, when Columbus sailed from Spain to North America, pricks of light begin to shine there. They increase in 1620 when the *Mayflower* arrives. A few decades later light illuminates the coast of South Africa when the Dutch land there. Next the British East India Company begins to send its people to India. In 1788 specks of lights begin to cluster on Australia's southeast coast when the British land there. Meanwhile, the sheet of light that has formed along the east coast of North America begins spreading westward. In 1836 a thin line of light arcs out from the sheet as pioneers begin to journey along the Oregon Trail.

The flow of movement over Erlich's light map illustrates migrations with which we are already familiar from written history. But remember, the data points that are the source of these glowing episodes were not culled directly from official documents but rather from family trees submitted by Geni's users. No doubt many of those trees are incomplete and contain some inaccuracies, but in the aggregate, a true history of the world emerges out of this collection of individual family histories.

The sociologist Wendy Roth said that in her field, one of the reasons that amateur research is regarded so dismissively is that it is considered "me-search": "It's research on 'me,'" she explained. "It's not really about broader themes and trends and theories; it's just about you and your particular little minute spot in history."

No doubt this is true of some genealogy. But with the massive digitization of all this family information (and, as will be discussed in chapter 8, DNA), it's becoming clear that in the aggregate there are extraordinary things to be said about collective genealogies.

The most ambitious genealogy in the world, at least the world's first very ambitious genealogy, is found in Iceland. Ingólfur Arnarson first settled the island in 874, after which some Icelanders sailed for distant shores and brought back slaves and concubines. The island was also visited by the occasional pirate or fisherman, but nevertheless the population has been relatively isolated, and today it is one of the more closely related populations in Europe. Icelanders have kept a remarkably detailed set of records since the beginning of their history. Some family chronicles date back as far as 1650, and in some cases to the eighth and ninth centuries. Genealogical information was recorded in historical documents, like the Book of Settlement, or Landnámabók, and the Family Sagas, or Íslendingasögur.

These ancient records have now been combined with church registers and censuses into an online database called Íslendingabók, or the Book of Icelanders. The total number of people born in Iceland since Norse settlement is about 1.3 million, and an utterly staggering half of these people are recorded in the Íslendingabók. (Iceland has had its share of plagues and volcanic eruptions; the living individuals recorded in the Íslendingabók represent only a subset of families from the early era that survived into the twenty-first century.) The director of the software company that assembled Íslendingabók, himself a family genealogist, said that digitizing all the records was like "working out a puzzle the size of a football stadium, with half the pieces missing and the rest randomly scattered."

When it first went online, a friend who lived in Reykjavík told me, it became a kind of dinner-party game, with everyone checking to see how they were related to one another. A Gallup poll in 2000 found that over 80 percent of Icelanders were enthusiastic about the project, but others felt there was something sinister in it. According to one op-ed, "Now a company in Iceland is recording in one place every piece of information documented in previously published works, including genealogies . . . and censuses. Unfortunately the company is doing this without asking anybody." Traditionally genealogical information has been in the public domain in Iceland, but increasingly it is being restricted. In 2013, however, some enterprising young Icelanders invented an app for mobile devices that could instantly calculate how closely related are, say, two people who met in a bar. If they bumped phones, the app would wirelessly match their identity in the Íslendingabók. The app's most talked-about feature was "incest prevention." "Accidentally sleeping with a relative has been a running joke in Icelandic culture for a while," one of the creators told the press. "Bump in the app before you bump in bed," they advised.

Large genealogies such as this are useful for learning about populations as well as individuals. In a similarly ambitious project in Canada, researchers used the genealogical data of more than one million Canadians to see if the choices made by different generations of European settlers had any impact on their descendants.

Quebec City was founded in 1608, and over the next ninety years the population steadily grew. Overall more than thirty thousand pioneering farmers from Europe settled northeastern Quebec before the end of the eighteenth century. Many of the earlier pioneers traveled from the city into the Canadian wilderness and established farms there. Others who arrived later tended to stay in the towns. To understand the forces that shaped the population, Damian Labuda and his colleagues used Quebec's extraordinary BALSAC population database, which includes official birth, death, and marriage records from 1680 to 1970. All of the individuals in the database are listed by the parish in which they lived— underlining the significance of religion as an organizer of human life— and all of the records are arranged in such a way that researchers could recreate the networks of family that existed over the three centuries that the database represents. They identified all the couples who married between 1686 and 1960 in the Saguenay-Lac-St-Jean region and then traced their descendants. Altogether, Labuda said, they looked at 1.8 million individuals and eighty-eight thousand marriages.

Labuda found out that the genetic rewards went to the bold: The pioneers who led the front wave of settlers, abandoning the more comfortable villages for the wild and establishing farms and families there, had more children than those who stayed behind in the cities. Not only did those first pioneers have more children, but their children also had more children, as did their grandchildren, and so on, which means that most of the current population of the Saguenay-Lac-St-Jean region can trace its roots back to the first wave of pioneer settlement, and not necessarily to the tens of thousands of otherwise perfectly healthy individuals who arrived some time later.

The progeny of the later arrivals have not disappeared from history altogether, though, as both the front and following waves contributed DNA to the current population. However, there were fewer of the latter, and their descendants had relatively fewer descendants. In fact, the front wave's genetic legacy was up to four times as large.

While it's often claimed that originally about nine thousand people founded the entire French Canadian population, this work reveals that there are patterns within patterns, and within all of French Canada some regions and some individuals from the founding population had a greater impact than their contemporaries because of their choices. While the population boom in all of Quebec was "spectacular," said Labuda, it was even more extraordinary in Saguenay, with a population of 10,000 people in 1850 and a population of 250,000 in 1950, a twenty-five-fold increase in one hundred years, mostly from local births. Given that a large part of the history of humanity, recorded or not, involves migrations into new territories, settlement, and then expansion outward again, the pattern may have important implications for groups in the past and the present.

The researchers found that the early women pioneers had 15 percent more children than the later ones; in addition, 20 percent more of their children married. Partly this may be due to the fact that women on the leading edge of the wave got married about a year earlier than the later arrivals. They bore children sooner, and they continued to bear children longer. It's possible that families who left the towns were not as exposed to the illness and disease that spread more easily in congested towns. In addition, Labuda speculates, in comparison to young people in the Old World, who had fewer children because there was much less to bequeath to them, young pioneers had more children simply because they could— there was more space to have a large family and more land to sustain them.

It's not clear how long this particular front-wave effect will last. Will it always be true that the population of the region has more DNA that can be traced back to the first pioneers than to those of any subsequent groups—will the DNA simply be recirculated throughout the area? Or do other huge events, subsequent migrations, or other factors have the power to change the pattern? Labuda's next project may throw more light on the issue. He and his colleagues plan to link multiple sets of histor- ical data to create a coherent genealogy for the entire historical popula- tion of Quebec from 1800 up to and including its present-day population of five million people. Similar projects are taking place on a smaller scale in parishes and villages in Finland, Italy, and Tunisia and in groups like the Amish in the United States and the Hutterites in Canada.

Population geneticists have long been interested in the patterns that arise when a species expands its range, but for the most part they have been able to study this only in animals that have a quick generational turnover. In the case of the Quebec study, the patterns would not have been discovered if people hadn't filled in their census forms—and in the case of the Geni history of the world, if they hadn't done their "me-search." But it's not just the records of families or the data gathered by genealogists that are now changing the way history and science are done; it's genealogists themselves.

Blurring the boundaries between family history, personal history, and social history is a project founded by University of Melbourne historian Janet McCalman. The goal of Founders and Survivors is to build de- tailed biographies of Tasmania's nineteenth-century convicts in order to chart the variety of paths they took and to discover patterns in the population as a whole. Despite the fact that generations of schoolchildren have studied Australia's convict past, McCalman explains that little was known about the vast majority of convicts after they left the system. "The ones we do know about were the exceptional people: the success stories, the winners. Until now, we have had little idea how many were losers in later life and whether there were effects or not, down through the generations."

Historians tend not to examine the fate of nineteenth-century populations exhaustively, partly because there aren't good enough records available through which to track them. McCalman focused on Tasmania's convict records because they are not only complete but also exceptionally detailed, including individual convicts' height, eye color, level of literacy, general disposition, and family background. Most important, the records have been made available online. Now anyone can sit back after dinner in the privacy of his or her home and—laptop at hand—unearth long-forgotten stories of hardship and adventure. McCalman capitalized on this by gathering a group of citizen-historians to crowdsource the past.

McCalman's volunteers, who each chose and researched a convict ship sent to Van Diemen's Land, are typically genealogists and descendants of the convicts. They try to find out what happened to every convict by trawling through census records in the United Kingdom, convict registers, old newspapers, and records of births, deaths, and marriages. Each ship is a "floating laboratory," and each group of convicts is a sample of humanity put through an extraordinary experiment in human resilience.

For the volunteers the experience is a bit like watching Michael Apted's groundbreaking documentary *Seven Up* series, but multiplied by tens of thousands of people over the course of more than a century. Retired Melbourne academic Garry McLoughlin didn't even know he was descended from a convict until he started volunteering with Founders and Survivors. "My great-grandfather was an early settler in Victoria, and we always knew there was something a little irregular about his origins," he explained.

In fact, McLoughlin was surprised to discover that his great-grandfather was innocent of the crime for which he had been transported. In 1853 Michael McLoughlin was convicted of stealing a gun, powder horn, and shot pouch from a local landowner in Dublin. Yet his alibi—that he was at the races at the time of the theft—was backed up by six witnesses. "It must have been terrible for him to have been transported—effectively for life," McLoughlin said. "But in the end his misfortune was my good fortune, as he began a family in Australia. If he'd stayed in Ireland, he might have died in the Great Famine, which began in 1845, just a year after he arrived here."

Leanne Goss, a stay-at-home mother, was drawn to the project because she had always been troubled by Australia's colonial past. Her research gave her a new empathy for the complexity of the era and for the people who were forced to leave everything that was familiar to them and travel to an unknown destination on the other side of the world with no hope of being able to return. "Some of them weren't nice, but most were just trying to survive. I've cried for some of them," she said. Goss's own ancestor, Samuel Marlow, came out on the *Godfrey Webster* in 1823. "Oh, he was a genuine criminal," she said. He stole plates from the London Mint.

So far the volunteers have been surprised to find how many convicts, once freed, remained broken by their experiences. Many did not marry or have children. But it's too early to draw definitive conclusions, Mc-Calman explained, as "half or more than half, especially after 1840, simply disappear."

Still, as the project progresses, the researchers will attempt to track convicts' descendants and perhaps understand how their lives and characters and the culture of silence around them influenced the fortunes of their offspring and their offspring's families down through many generations. No one is expecting a deterministic connection—as with the Canadian frontier, where the ongoing fertility of a family was influenced by whether it had an ancestor in the first wave, any effects would be probabilistic. If certain experiences—education, for example, or the presence of a mother, or the length of servitude—significantly helped people change their fate in such tough circumstances, the project may uncover them.

Many of the tales are, by their nature, high-stakes historical dramas. David Noakes followed the travels of the convict William Anthill from Tasmania to New Zealand and beyond. In New Zealand, Anthill, who was born in 1823 in Leicestershire, boarded a ship to England called the *Blue Jacket*. Three hundred miles south of the Falklands, the ship caught fire, and the captain and passengers headed off in one lifeboat, while the crew took two others. Each boat brought a box of gold for ballast. Anthill's lifeboat floated in the Atlantic for three weeks. Three men died, and he and his fellow survivors were forced to kill the ship's dog and drink its blood. An article from the *Times* of London later reported that the desperate men opened the gold boxes and sucked on the "ingots in the same manner that men suck on pebbles in an attempt to slake their thirst." When they were finally found, the lifeboat was scattered with gold and spattered with blood, and their rescuers assumed Anthill and his party had murdered the *Blue Jacket*'s crew to steal the gold. The starving survivors were clapped in irons and released only when word came from the rescued captain in England that they were innocent. Anthill eventually returned to New Zealand, where he raised a family and was later sued for bigamy. He died in 1902. Noakes remarked in response to encountering such histories, "I hardly read novels anymore."

In addition to information, families also pass down beliefs and behaviors—at least, we believe we know this, because it seems so obvious. Gisela Heidenreich taught her clients family-systems therapy, which is based on the idea that emotional predispositions and behavior may be transmitted through generations. Family-systems therapy, which was developed from clinical experience, has been very successful in many countries. Until recently, however, no one had ever tried to systematically measure whether and how culture is transmitted from one generation to another and, if indeed it is, to determine for how many generations it is passed along.

Chapter 7

Ideas and Feelings

They fuck you up, your mum and dad
They may not mean to, but they do
They fill you with the faults they had
And add some extra, just for you
                        —Philip Larkin, "This Be the Verse"

When Olaudah Equiano was eleven years old, the adults in his eighteenth-century village would head off in the morning to work in the fields. The children who remained behind would meet up in someone's house to play, and whenever they did, at least one of them would take up a post in a nearby tree to keep a watch out for strangers. One day, Equiano wrote, he and his sister were home alone with no lookout when two men and a woman suddenly appeared over the walls of their compound. With no time to scream, the children were seized and their mouths covered as the trespassers took them back over the wall.

The kidnappers journeyed to a waiting place, and along the way Equiano spied people in the distance. He cried out for help, but his captors gagged him and put him in a sack. That night he and his sister clung to each other, but the following morning his sister was taken away, never to be seen again. "The small relief which her presence gave me from pain was gone," he later wrote. "And the wretchedness of my situation was redoubled by my anxiety after her fate, and my apprehensions lest her sufferings should be greater than mine." Of his sister he wrote: "Though you were early forced from my arms, your image has always been rivetted in my heart, from which neither time nor fortune has been able to remove it."

It is true that the convict system destroyed families and communities and caused enormous suffering: It lasted from the early seventeenth century to the late nineteenth, during which time it is estimated that two and a quarter million convicts were transported, and even though many ex-convicts survived and went on to prosper, many did not. But even so, the period in which convicts were subjected to forced labor overlapped with one that saw a system that was even more cruel and lasted much longer: slavery.

The slave trade in Africa took place over many hundreds of years and in fact consisted of four different slave trades. From as early as the ninth century the trans-Saharan slave trade abducted people and brought them to northern Africa; the Red Sea slave trade and the Indian Ocean slave trade transported them to the Middle East, India, and Indian Ocean plantation islands. The Atlantic slave trade primarily took slaves to the New World. In many African countries at least twenty generations lived and died in a world in which a spouse or a friend or a child might suddenly disappear without a trace. Many Africans believed they were being captured and shipped over the ocean to be eaten.

The insecurity of life in a world of slavery is hard to imagine, let alone the extraordinary length of time that the threat of abduction loomed. Some eighteenth-century Africans, like the members of Olaudah Equiano's village, began to leave children in locked stockades under armed guard as they went off to work in the fields.

But the threat didn't always come from outside. In the course of a conflict one village or ethnic group might attack another, and sometimes entire villages would be surrounded by horsemen and burned down; those who did not die in the initial attack were tied together and taken away. Communities also turned on themselves. Farmers who were desperate to defend themselves against attack abducted other villagers and sold them to slave traders so they could afford iron knives or firearms. As local communities collapsed, even their chiefs became traders, offering their people up as tributes, sometimes hundreds of them in a single year. In some areas all the traditional punishments for lawbreakers evolved into the single sentence of being sold to slavers, and people were often falsely convicted of crimes from adultery to witchcraft in order to supply merchandise to the trade. The chief of the Cassanga tribe made accused criminals drink a poisonous red liquid in the "red water ordeal." Those who vomited were declared guilty and sold as slaves. Those who didn't vomit died—and their family were sold as slaves.

Even worse, villagers found themselves tricked and betrayed not just by close neighbors but by family members in order to pay off debt. The nineteenth-century German missionary Sigismund Koelle asked over 140 ex-slaves how they had been taken. Almost 20 percent of them told him that family or friends had given them up. An anthropologist who stayed with the Kabre of Togo in the 1990s said they have "surprisingly vivid memories" of the period, and they were matter-of-fact about it. Kabre locals pointed out the houses where people who sold their relatives had once lived, and many of the names of people who sold their kin were remembered. Often the seller was a man who, according to tradition, had ownership rights over his sister's young children. Even today Kabre uncles may jokingly threaten their nephews that they will sell them to slavers.

It may be that more than thirty million Africans were wrested from their homes and families, and the great majority never saw either again. In just a single period of the Atlantic slave trade, between 1700 and 1850, ten million people were shipped across the ocean. The diary of a sailor from this era records the anguish of the slaves on a ship that was about to leave Africa: "The slaves all night in a turmoil. . . . They felt the ship's movement. A worse howling I never did hear, like the poor mad souls in Bedlam Hospital. The men shook their fetters which was deafening."

Of the ten million who were shipped, only 8.8 million made it to the other side of the ocean, many lost their lives along the way.

Today the continent is afflicted by many problems related to underdevelopment, much of which has been attributed to the legacy of the slave trade and also to the colonial period, which lasted from 1885 to 1960. While historians have made a compelling argument about the long-term damage of slavery and colonialism using extensive documentation from the period, until recently no one had attempted to quantify that damage. As a graduate student Nathan Nunn, now a Harvard economist, began to compare different economies in modern Africa, and he found that the countries that lost more people to the slave trade were also the poorest countries today.

How could the slave trade shape economies and affect lives in Africa over a century after it ended? Nunn discovered that the legacy was passed on not only in the materials and institutions from the past but also in the way people thought about one another.

Nathan Nunn was born in a log cabin on a Canadian ranch that was so remote it could be reached only by plane or snowmobile. One of his earliest memories is of forlornly looking through a gap between the floor logs where his toy truck had fallen. Nunn lived with his American father and South Korean mother, who were there because his father had migrated from Montana to Canada to try to breed a hybrid of a cow and a yak, an animal that would tolerate minus-forty-degree temperatures better than the typical cow. Years earlier he had met Nunn's mother in Seoul. They corresponded, and then she flew to the Canadian west coast to marry him. The move from a city of seven million people in humid Southeast Asia to a ranch with ten people in the Canadian tundra was enormous. She told the young Nathan stories about the culture shock, like the time when he was very little and the family's horse-drawn sleigh crossed a frozen lake. The ice broke, the horses fell through, and Nunn's father had to cut them from the ice and chop down a bunch of trees on the shore to get a fire going so they wouldn't freeze.

On a cold New England day I met Nunn outside at his Harvard office. He had a boyish face and a casual air that belied the years he has spent digging through archives to amass records for one hundred thousand slaves, as well as his reputation among economists as a trailblazer. We sat down and spoke about the systematic differences between cultures, how history matters, and why the countries that lost more people to the slave trade were also the poorest countries today.

In order to find a connection between slavery and modern economies, Nunn asked if the differences in economic well-being today could be explained by differences that existed before the slave trade. Were the countries that were already poor the same countries that were more engaged in the slave trade? In fact, Nunn found the opposite: Regions that lost the most people to slavery had once been among the best-developed economies and best-organized states on the continent, with central governments, national currencies, and established trade networks. It was the states that were least developed and had higher degrees of violence and hostility at the time of the slave trade that were better able to repel slavers and not suffer the long-term effects of the trade.

Could the relationship between modern poverty and historical slavery be explained by the subsequent effects of colonialism or by the natural resources possessed by a country? Nunn found that although those factors appeared to have an effect, neither was as powerful. It was slavery that mattered, and it mattered greatly.

When he was a graduate student, Nunn read about Olaudah Equiano's life and Sigismund Koelle's account of slavery and was stunned by the number of first-person reports about friends and family selling someone into slavery: Almost 20 percent of slaves had been betrayed by people to whom they were close. He began to wonder what kind of long-term impact such betrayal might have. Then one day he met someone who had been asking the same question but for different reasons. Nunn gave a presentation about his slave data, and afterward a man named Leonard Wantchekon introduced himself. Wantchekon grew up in Benin, one of the countries that was most impacted by the slave trade. Nunn's ideas about how the trade affected modern economies deeply resonated with Wantchekon's own experience, who believed that trust was an important part of the story.

Wantchekon had been thinking about trust for a long time. As a young student he was dedicated and bright and particularly talented in math, but in college in the 1980s he became more and more involved in political activism. He organized protests and distributed leaflets criticizing Benin's brutal government. When some of his friends were arrested, he had to go into hiding. For five years Wantchekon changed his location every other day, sometimes sleeping in a cave or a forest. Later he was arrested and tortured. His guards made him stand up for three days and nights and then beat him for hours at a time. After eighteen months Wantchekon escaped jail, fleeing to Nigeria and from there to Canada.

Despite his experiences at the hands of the guards of Benin's Petit Palais, as the intelligence headquarters was known, Wantchekon told me that the worst of Benin was not the overt corruption but the distrust between people who were closely connected. That suspicion existed everywhere: in economic activity, in political activity, and in family life. When he was a student, people would turn on each other for no apparent reason. Friends would accuse each other of being witches. Distrust was evident in proverbs like "You can escape your enemies, but not your neighbors and family members. So beware of those you know." It featured in popular songs, which had lyrics like "This guy, he looks good, but be careful, he can hurt you" or which would explain who was trustworthy (a brother from the same mother) and who wasn't (brothers from different mothers, cousins, other relatives). All his life Wantchekon's mother had warned him to be wary of his great-aunt Awetinjo, lest she bring him harm, and yet his mother demonstrably cared for the old woman. After his mother died and he started to live his life on the run, Wantchekon knocked on the old lady's door. She was on her deathbed but got up to greet him. Her only thought was to say something kind: "When I get to heaven, I will meet your mother and tell her not to worry, that you are okay."

Distrust was evident even in the language of small children. As Wantchekon recalled, if kids of nine or ten warned each other off someone, they would say, "He can sell you" or "He can make you disappear." It wasn't until Wantchekon left Benin that he even questioned the literal meaning of the phrase "sell you." It must have been a remnant of the slave trade.

Canada gave Wantchekon refugee status, and before too long he completed an undergraduate degree and then went on to do a PhD in economics. He was made a professor of economics at NYU and later Princeton. After their first meeting, he and Nunn began to work together.

They began with the intuition that trust could be a channel through which slavery still affects modern economies. But their goal was to find evidence for it. Of course, trust is a crucial part of any economy: Societies must have some degree of trust in order to be able to trade. At the most basic level, if people don't trust one another, they are less willing to take a chance in business, whether it involves a simple exchange of goods or a complicated contract. But no one in economics had ever tried to measure the relationships among history, trust, and the economy before. After all, trust was an element of culture, and "culture" was a vague, fuzzy concept. Nunn and Wantchekon defined it as simply as they could: Culture, for their purposes, was the rules of thumb people used to make decisions. *Do I trust this person? Do I distrust him?* People from different cultures use different rules of thumb to make such determinations.

Building on Nunn's finding that the countries that lost more of their populations to the slave trade over one hundred years ago were also the poorest today, Nunn and Wantchekon examined the Afrobarometer, a survey project that measures public attitudes to different aspects of African daily life, like democracy, employment, and the future of citizenship. It is comparable to a Gallup poll, and it includes seventeen countries. The researchers found that overall, people tended to have more trust in those who were closer to them—for example, friends over government officials. This was a universal pattern. But it was also the case that the groups that were most exposed to the slave trade over one hundred years ago were also the groups with the lowest levels of trust today. Modern Africans whose ancestors lost the most people to slavers distrusted not just their local government and other members of their ethnicity but also relatives and neighbors much more than Africans whose ancestors were not as exposed to the slave trade.

Did the slave trade give rise to a culture of mistrust that was passed down from the slave era even to individuals who live in the same places today? There are good reasons to believe that it might have. For those who witnessed the ways an innocent bystander might be swept up by or somehow betrayed into the slave trade, it would have made more sense to distrust people, as a general rule. People who automatically distrusted others were probably more likely to do well, or at least to not be enslaved. Wariness would also have been a smart strategy to teach the next generation.

There's another way this terrible correlation could be interpreted: Perhaps the slave trade made people not less trusting but less trustworthy. Perhaps people weren't trusted in countries like Benin because they didn't deserve to be trusted. After all, chiefs turned on their own people, and families sent some of their own literally down the river. Was a culture of betrayal passed down as well as a culture of distrust? This could partially be the case. Nunn's analysis reveals that ethnic groups and local governments in the regions that were most affected by the slave trade in the past are also least trusted today. People whose ancestors were more affected by the slave trade were more likely to report that they did not approve of their local councilors, who were corrupt and did not listen to constituents. As Nunn explained, it's quite likely that this is an accurate assessment of the local councils in these areas. Nevertheless, when they controlled for this effect, there was still a significant amount of *distrust* in countries most affected by the slave trade—regardless of whether the object of trust was truly worthy.

When Nunn and Wantchekon published their study, Wantchekon spoke about it on a television show in Benin, and it struck a very deep chord. Many locals wrote to him, and it seemed that everyone had something to say about it. One old friend phoned and put the call on speakerphone so that Wantchekon could hear his entire family excitedly affirming and arguing about the idea of trust. It was as if a fever had broken. Everyone acknowledged that a deep distrust still shadowed their lives, and they also agreed that being so suspicious made no sense. They *should* trust one another more.

Still, even years after his study was published and after he had lived in the United States for a long time, Wantchekon rang his sister in Benin to tell her that an old friend was visiting him. She warned, "You know, you should be careful. Watch out for him!" Wantchekon thought, *I've known this guy for forty years! He hasn't killed me in forty years!* Trust remains a major topic in public meetings in Benin. Indeed, Wantchekon has now started a university in the country. "The best contribution that I can make to Africa and to Benin in particular is by training the next leaders and the next academics through a strong graduate program in economics," he said. "In September we are going to start, everything is on track."

Nunn and Wantchekon's assertion that mistrust and silence could be passed down for more than a century was shocking. We don't normally think of ideas and attitudes persisting for so long. Could they be passed down over even longer periods of time?

In 1348 in a castle on Lake Geneva, a Jewish man named Agimet was tortured "in the presence of a great many trustworthy persons." Eventually he broke and confessed to having caused the Black Death by poisoning the local wells. In the previous year the plague had swept into Europe via the Silk Road. In village after village in Europe, people awoke to find themselves feverish, the skin of their fingers and toes turning black, and their lymph nodes swelling grotesquely until they split open and bled. Victims bled internally as well, urinating blood and coughing it up, until they died in great pain.

The plague, which was highly contagious and destroyed entire families and villages, is believed to have killed as many as fifty million people in Europe (60 percent of the population) and seventy-five million worldwide. At the time, no one understood what the dread affliction was or where it came from. Some believed it must be a punishment from God, a consequence of the movement of the planets, or a malady created by humans. Many people blamed it on the Jews, the largest minority in Europe at the time, or on people with disabilities, and in some rare cases even on the nobility. Primarily, though, the Jews were held accountable, and after Agimet's torture, terrible pogroms (violent riots that target a specific ethnic group) took place all over Europe for more than a decade.

Out of 320 towns that had a Jewish community in the territory that later became Germany, 232 carried out pogroms, destroying homes, inflicting torture, and expelling or killing Jewish inhabitants. In many areas entire communities were disbanded, and fleeing Jews were set upon by peasant mobs. Seventy-nine towns remained peaceful. The Jews even had defenders among the Christians. Pope Clement VI declared that well poisoning was a crime "without plausibility." Medical faculties in many towns also asserted that the stories about well poisoning were false. But reason had little effect on the panic. As the plague took hold in Basel, Switzerland, and more Christians died from the infection than Jews, on January 9, 1349, some six hundred Jews were forced into a specially constructed wooden building on an island in the Rhine, where they were burned alive.

After three years the worst of the plague was over, but the fear and

the hatred it engendered persisted. In one of the most remarkable studies of the transmission of ideas over time, the economists Nico Voigtländer and Hans-Joachim Voth found evidence that animosity endured
generation after generation, for as long as six hundred years.

Voigtländer and Voth compared the treatment of Jewish people in
towns after the Black Death with their treatment in the same towns in
the 1920s. Following World War I, anti-Semitism was on the rise in Germany. Many Germans blamed Jews for the war, and once again villagers
turned on their own neighbors, carrying out pogroms. The researchers
identified the towns that exhibited the most virulent anti-Semitism in pre–
World War II Germany, and they found a remarkable correlation with
the Black Death pogroms. Of the twenty pogroms that took place in the
1920s, Voigtländer and Voth found that nineteen were carried out in
towns that had also attacked their Jewish communities in the fourteenth
century. If you were a Jew in 1920s Germany and you lived in a town
where no medieval pogroms took place, your chance of being attacked
by your fellow townsfolk was 1.1 percent. But if you were Jewish and you
lived in a town where a medieval pogrom had occurred, the chance that
you would be attacked rose to 8.2 percent.

The researchers compared the cities of Aachen and Würzburg, which
were similar in size before World War II. Jews had lived in Würzburg
since 1100, while Aachen had had a Jewish community since 1242.
There is no record of any violence against Jews in Aachen before or during the Black Death. By contrast, the citizens of Würzburg turned on
their Jewish community and killed eight hundred people. Voigtländer
and Voth noted the sentiments of medieval Würzburg's notary, who wrote
to his bishop, "The Jews deserved to be swallowed up in flames." Over
six hundred years later, even though both communities destroyed their
Jewish synagogues, only Würzburg had pogroms.

Voigtländer and Voth didn't only examine direct violence against Jews
in 1920s Germany; they tracked anti-Semitism in a number of ways.
One measure was the poll performance of the Nazi Party in 1928, when
the Nazis did not yet have mass popularity. "In places with a history of

Jew-burning," wrote Voigtländer and Voth, "the Nazi Party received 1.5 times as many votes as in places without it."

Letters to the editor of *Der Stürmer*, a particularly racist Nazi newspaper, also showed a link with the very distant past. The researchers determined the location of writers of this correspondence and found a strong link with towns that had carried out medieval attacks on Jews. In the 1920s residents of Würzburg wrote ten times as many anti-Semitic letters to the editor as did residents of other towns.

"Dear *Stürmer*," wrote one schoolgirl. "Regrettably, [the students in my school] still have many Jewish fellow students. Equally regrettably, many German girls are still close friends with these Jewish girls. . . . I consider these friendships very dangerous since the Jews and their corrupting ideas destroy the souls of the girls slowly but surely."

After 1939 more Jews were deported to camps from areas that had a history of medieval violence than from areas that did not. Even though deportation during this period was a national policy, Voigtländer and Voth argue that rules were enforced by local authority: The numbers of Jews who were deported reflected how stringently local administrators judged their citizens' proof of ancestry. Eric Ehrenreich came to the same conclusion when he examined genealogy and ancestral proof. Even during *Kristallnacht*, when all across Germany the hateful treatment of Jewish people was effectively licensed by the Nazi Party, more Jewish synagogues were attacked in towns where Jews had been killed six hundred years earlier than in towns where they hadn't.

Voigtländer and Voth are not suggesting that relationships among different cultures in Europe were always harmonious before the Black Death. In fact, pogroms took place before the 1300s. (In England, where there were no pogroms following the Black Death, that was due not to the absence of hatred so much as to the fact that the English had already expelled their Jewish population in 1290). Still, many more pogroms occurred in Europe after the Black Death hit.

The details of medieval pogroms and twentieth-century anti-Semitic attacks have been known for a long time, but until Voigtländer and Voth's analysis, no one had tried to determine whether the two might be connected. Partly this is because no one had imagined that attitudes could endure for so long.

If the connection between hateful acts across the centuries and between trauma and distrust over generational time is real, how do the ideas and the feelings persist? How might they be passed down from one generation to the next? Can the personal qualities of a great-great-great-grandfather, like his faith in people or his suspicion of them, really influence the feelings of his descendants today?

I asked Voigtländer how such hatred was preserved over the course of centuries. He and Voth found that in cities that grew significantly after 1750, the long-term transmission of anti-Semitism was disrupted. Crucially, these rapidly industrializing cities expanded because many people moved in from elsewhere, not because the locals had more children. Where Jew-hatred persisted, there were relatively fewer people coming in. "Even though migration everywhere increased rapidly after 1820, most inhabitants of a typical town in our sample must have been direct descendants of those who lived there in 1350," they wrote.

Nunn also wondered about who was passing down the distrust in Africa. There was no way to track it through time beyond the records that he and Wantchekon had already examined. There was also no literature that surveyed trust in earlier periods, no handy Afrobarometer that logged attitudes in different families and communities over time.

So Nunn and Wantchekon estimated for each individual whose trust was surveyed how many slaves were taken from his personal ethnic group and how many slaves were taken from the areas where he lived. The underlying idea was that, if you are taught values primarily by your family, you are likely to take those with you wherever you go. But if you absorb values from the people who live around you and from the legal, social, and political institutions in a particular region, then where you live may have more impact on your beliefs than what you learned from your parents. If this was the case, then when someone left his native village and moved to a place with a different predominant attitude, he might modify his existing values by virtue of being surrounded by people with different ones. Similarly, if someone moved into a region where all the civic organizations seemed to foster distrust, then even if he came from a previously trusting place, he might take up the local culture and become less trusting himself.

It seemed that both families and social institutions matter but that the former is more powerful. The data suggested that a region might develop its own culture of distrust and that it could affect people who moved into that area, even if their ancestors had not been exposed to the historical event that destroyed trust in the first place. But if someone's ancestors had significant exposure to the slave trade, then even if he moved away from the area where he was born to an area where there was no general culture of mistrust, he was still less likely to be trusting. Indeed, Nunn and Wantchekon found evidence that the inheritance of distrust within a family was *twice as powerful* as the distrust that is passed down in a community.

This accords well with our personal intuitions about families: The people who raise us shape us, intentionally or unintentionally. The people who raise us were likewise shaped by the people who raised them, and so on. Similarly, the way we treat other people, even our offspring, is shaped by the way we were shaped. This is not to say that our peers don't affect our attitudes, nor does it mean that the society in which we choose to live doesn't contribute as well. Obviously, the older we get, the more we develop the ability to shape ourselves. Family history doesn't necessarily determine who we become, but this body of work suggests that the effect of a family may be so powerful that it can be replicated down through many generations, over and over through hundreds of years. It's no wonder that so many people choose to study the distant histories of their families to understand how they work today. If genealogists believe there isn't enough in their daily lives or their culture that sufficiently explains who they are—either to others or to themselves—it may be because they are right.

In fact, the legacy of a family may be so powerful that it will not only last over extraordinary periods of time but extend over great distances as well.

Catastrophic events like the plague or slavery are not the only ones that echo down the generations. Widespread and deeply held beliefs can be traced to apparently benign events too, like the invention of technology. In the 1970s the Danish economist Ester Boserup argued that the invention of the plow transformed the way men and women viewed themselves. Boserup's idea was that because the device changed how farming communities labored, it also changed how people thought about labor itself and about who should be responsible for it.

The main farming technology that existed when the plow was introduced was shifting cultivation. Using a plow takes a lot of upper-body strength and manual power, whereas shifting cultivation relies on hand-held tools like hoes and does not require as much strength. As communities took up the plow, it was most effectively used by stronger individuals, and these were most often men. In societies that used shifting cultivation, both men and women used the technology. Of course, the plow was invented not to exclude women but to make cultivation faster and easier in areas where crops like wheat, barley, and teff were grown over large, flat tracts of land in deep soil. Communities living where sorghum and millet grew best—typically in rocky soil—continued to use the hoe. Boserup believed that after the plow forced specialization of labor, with men in the field and women remaining in the home, people formed the belief—after the fact—that this arrangement was how it *should be* and that women were best suited to home life.

Boserup made a solid historical argument, but no one had tried to measure whether beliefs about innate differences between men and women across the world could really be mapped according to whether their ancestors had used the plow. Nathan Nunn read Boserup's ideas in graduate school, and ten years later he and some colleagues decided to test them.

Once again Nunn searched for ways to measure the Old World against the new. He and his colleagues divided societies up according to whether they used the plow or shifting cultivation. They gathered current data about male and female lives, including how much women in different societies worked in public versus how much they worked in the home, how often they owned companies, and the degree to which they participated in politics. They also measured public attitudes by comparing responses to statements in the World Value Survey like "When jobs are scarce, men should have more right to a job than a woman."

Nunn found that if you asked an individual whose ancestors grew wheat about his beliefs regarding women's place, it was much more likely that his notion of gender equality would be weaker than that of someone whose ancestors had grown sorghum or millet. Where the plow was used there was greater gender inequality and women were less common in the workforce. This was true even in contemporary societies in which most of the subjects would never even have seen a plow, much less used one, and in societies where plows today are fully mechanized to the point that a child of either gender would be capable of operating one.

Similar research in the cultural inheritance of psychology has explored the difference between cultures in the West and the East. Many studies have found evidence for more individualistic, analytic ways of thought in the West and more interdependent and holistic conceptions of the self and cooperation in the East. But in 2014 a team of psychologists investigated these differences in populations within China based on whether the culture in question traditionally grew wheat or rice. Comparing cultures within China rather than between the East and West enabled the researchers to remove many confounding factors, like religion and language.

Participants underwent a series of tests in which they paired two of three pictures. In previous studies the way a dog, a rabbit, and a carrot were paired differed according to whether the subject was from the West or the East. The Eastern subjects tended to pair the rabbit with a carrot, which was thought to be the more holistic, relational solution. The Western subjects paired the dog and the rabbit, which is more analytic because the animals belong in the same category. In another test subjects drew pictures of themselves and their friends. Previous studies had shown that westerners drew themselves larger than their friends. Another test surveyed how likely people were to privilege friends over strangers; typically Eastern cultures score higher on this measure.

In all the tests the researchers found that, independent of a community's wealth or its exposure to pathogens or to other cultures, the people whose ancestors grew rice were much more relational in their thinking than the people whose ancestors were wheat growers. Other measures pointed at differences between the two groups. For example, people from a wheat-growing culture divorced significantly more often than people from a rice-growing culture, a pattern that echoes the difference in divorce rates between the West and the East. The findings were true for people who live in rice and wheat communities today regardless of their occupation; even when subjects had nothing to do with the production of crops, they still inherited the cultural predispositions of their farming forebears.

The differences between the cultures are attributed to the different demands of the two kinds of agriculture. Rice farming depends on complicated irrigation and the cooperation of farmers around the use of water. It also requires twice the amount of labor that is necessary for wheat, so rice-growing communities often stagger the planting of crops in order that all their members can help with the harvest. Wheat farming, by contrast, doesn't need complicated irrigation or systems of cooperation among growers.

The implication of these studies is that the way we see the world and act in it—whether the end result is gender inequality or trusting strangers—is significantly shaped by internal beliefs and norms that have been passed down in families and small communities. It seems that these norms are even taken with an individual when he moves to another country. But how might history have such a powerful impact on families, even when they have moved away from the place where that history, whatever it was, took place?

How do immigrants reproduce old values once they have left behind the old institutions and local beliefs that reinforced them? Raquel Fernandez and Alessandra Fogli wondered if second-generation daughters of immigrants to the United States might still be affected by the values of their parents' home countries, even if the women themselves had never been there. Their subject group was women in the 1970s who were born in the States but whose parents came from elsewhere, and they asked specifically how much they worked and how many children they had.

If the contemporary U.S. society in which the women grew up had the biggest effect on their lives, argued Fernandez and Fogli, then their work and family profiles should resemble those of their American peers who were not children of immigrants. If the culture of the family was a more significant factor, then women's choices would look more like those of their grandparents' generation. The researchers compared data about work and children in the 1970 U.S. census with data from each woman's parents' country of origin in the 1950s. They found that even when the present mattered, the past still had significant influence. If the 1950 data revealed that women in the parents' country of origin worked more, then the U.S.-born daughter worked on average a week more every year. If the 1950s data showed that women in the parents' country of origin had more children, then the U.S.-born daughter had more children than her peers who were not second-generation immigrants.

Fernandez and Fogli controlled for the effects of the husband's and wife's education, the education of their parents, the income of the husband, and the geography of markets. Even taking into consideration all these factors, it looked like the attitudes of the Old World still shaped the choices of women born in the new.

The researchers also asked if the amount that a woman worked and the number of her children were more powerfully shaped by her own culture or her husband's. Apparently, when a wife's and husband's parents' cultures of origin are different, the husband's parents' culture of origin was the more significant factor. It's unclear why this would be so, but as Fernandez and Fogli observe, the choice of marriage partner itself is hardly random.

It is possible that the interaction between a family and a small community makes the family's force even more powerful, suggest Fernandez and Fogli. When there is no clear separation between the people at home and the people in the neighborhood schools and churches and other institutions, the researchers found that the power of a family appeared to be greatly enhanced. The parents'-country-of-origin effect on women was magnified when the U.S. women were raised in communities surrounded by many other families from their parents' ethnic group. The higher the percentage of the parents' ethnic group in the neighborhood, the more likely it was that the modern women made choices that were influenced by the old ways. Even though they were isolated from the institutions of the Old World, they were still surrounded by people from that world. The groups kept the old beliefs alive and passed them on.

No study has found a single universal principle that dictates how beliefs and attitudes are reproduced down the generations. One study found that, like the second-generation immigrant women of the 1970s United States, Irish Americans in the 1910s had fewer children than their peers in Ireland but still significantly more than their peers in the States. By contrast, the childbearing patterns of German immigrants to the United States at the time showed no connection whatsoever with the culture of childbearing back in the old country.

Some cultures seem to perpetuate a community closeness that in turn fosters a perpetuation of values. Immigrants from Mexico, Italy, and Japan are more likely to cluster together in new neighborhoods and presumably to maintain the historical beliefs that shaped them. The Turkish, French, and Lebanese, by contrast, are less likely to live in a neighborhood with many people of the same ethnicity.

The way a trait is passed down in a culture may also depend on the trait itself. Trust affects economies all over the world, but in Italy, for example, it appears to have been shaped by different forces than in Africa. Guido Tabellini examined trust, respect for others, and "confidence in the link between individual effort and economic success" in economies in southern and northern Italy by comparing answers to questions like "[Would] you say that most people can't be trusted or that you can't be too careful in dealing with people?" Even though all the regions he examined had effectively the same contemporary level of literacy and the same quality of institutions, some of them had been less literate and had more institutional corruption in the past. Tabellini found that these latter regions had less trust, respect, and confidence today, as well as poorer economies.

As Nunn was beginning his research, his field was undergoing something of a revolution. A group of economists had begun to explore the way that history could influence an economy. Obviously, an economy may be affected by such immediate factors as the destruction of important institutions, the death of key figures, the failure of crops, and the spread of disease, and the more recent such an event was (such as the terrorist attacks on America on September 11, 2001), the easier it was to assess its economic impact. But now economists were setting out to try to measure the impact of distant historical events *through* time. They began to talk about *horizontal transmission*, the things that are learned from one's peers and society, and contrast it with *vertical transmission*, meaning, essentially, what gets passed down. The idea inspired an enormous amount of work on the impact of colonialism, especially the way that institutions like banks, governments, and the legal system were shaped by colonialism in different countries and the way that they in turn later affected their countries' economies. It was the first time that economists made an evidence-based case that history mattered. Their work was particularly fruitful because it wasn't based just on this general notion but also provided a way to connect specific outcomes to particular events.

But even though the idea that history could be measured was being taken seriously, most of the work focused on the impact of social institutions. Asking how hate or fear might affect the well-being of an economy was still considered unscientific. Partly this was a reflection of the lack of access to data that revealed people's beliefs and the difficulty of defining precisely what culture is. This is now changing as more information about beliefs and attitudes becomes available, as large amounts of data from the past become easier to handle, and as researchers come up with innovative ways to interpret it. But partly—and ironically—researchers did not examine the economic consequences of elements like trust and hate and fear because of a *belief* that these things didn't matter beyond the lives and lifetimes of individuals.

Yet as Fernandez and Fogli point out, markets have a fundamental relationship with beliefs. A culture's belief about the permissibility of selling another human being as chattel will affect whether it has a slave trade and how widely it operates. The belief that it's good for women to work outside the home will affect the size of the workforce. Since their research was published, Wantchekon said, the dismissal of culture as a factor in studying economics has changed: "'Culture' was no longer a dirty word."

What lessons might we take from these extraordinary connections between ancestral experience and modern attitudes? First, let's be clear:

No one is suggesting that the lives of our ancestors may be examined like a fortune-teller's deck of cards and our own fate foretold from them. Correlation is not causation. Our forebears may simply have nothing to do with our psychological makeup at all. But we may in some respects be profoundly shaped by what happened to those who came before us, and sometimes the past matters, whether we are actually aware of it or not.

It strikes me that being cognizant of the grand historical arcs our families have lived through could also enable us to better see what qualities we have freely chosen for ourselves and what we have unthinkingly inherited from our great- and many-times-great-grandparents who lived in a different time and possibly wanted quite different things from their lives. If someone discovers that one or many of his ancestors were immigrants or exposed to the slave trade or a plague or famine, he may find that that knowledge illuminates some aspect of his life now, whether it's an idiosyncratic word his father uses, his own reluctance to travel, the number of his own children, or his family's penchant for not talking about family.

Historians, of course, have been telling us for hundreds of years that history matters and that we as a society cannot be free from the past if we don't learn from it. Now the fascinating correlations found in this economic research suggest that distant historical events may influence the character of a modern family and that the choices of families can illuminate big history. Recall Ralph Waldo Emerson's cry, "Why should not we also enjoy an original relation to the universe?" Maybe we can, but it will surely help if we can identify what was passed down to us and what we have freely chosen for ourselves.

Obviously, the circumstances of our lives shape us too. Education has a huge impact, as do job opportunities. Personal income can, of course, change everything. The influence of particular individuals, whether teachers, mentors, or spouses, also matters.

All these factors may interact with one another as well. It's complicated and recursive. We are shaped by events, and then we shape people who initiate other events. We are shaped directly by people, and we shape others accordingly. Documents and ideas and feelings that are passed down may tell us about this; DNA is a record, and it is passed down too. What can it tell us?

Chapter 8

The Small Grains of History

You may not be able to leave your children a great inheritance, but day by day, you may be weaving coats for them which they will wear for all eternity.

—Theodore L. Cuyler

Westray is an hour's ferry ride from Mainland through the dark black water of the North Sea. The island of Mainland is itself an hour's trip from the coast of Britain, which is, of course, itself a relatively small island off the western coast of Europe. Of the tens of millions of people who were either born on or swept into the British Isles over the last ten thousand years, there are six hundred people left in Westray, and only ten thousand in the Orkney island group, of which it is a part.

The road from the ferry terminal to Pierowall, Westray's biggest town, rolls up and down the length of the island, traveling through pale green fields and sections of craggy rock, at points opening out to reveal vistas of the sea at either side. Along the eastern cliffs puffins are drawn with such precise lines they look prim against the wildness; on the beach fat selkies loll. Their velvet hides and anime eyes almost distract from the spectacle they create when they flop toward the sea—a reminder that evolution does not make überathletes for every niche but only does enough to get by. Even in May the Arctic wind has an icy hand.

Legend has it that a Spanish galleon sank off Westray's coast in 1588. Sailors swam to the islands, and those who weren't dashed on the rocky spires were welcomed. But on Westray's neighbor, Papa Westray, it soon became clear there wasn't enough food for the winter for everyone, so the locals pushed the poor sailors over the cliffs until there were none left.

On Westray, though, the ones who made it safely to shore proved helpful enough to keep. They married local girls, and the accidental im-migrants and their descendants were thereafter known as the Dons. They were renowned as extroverted performers and great sailors, yet after the first generation the Dons kept to themselves. Romance with the locals was forbidden, and one young Don who broke the rule, so the story goes, was murdered by his cousins. The Dons would have made a striking contrast to the pale, light-eyed locals. For many years people on the island who had dark hair or olive skin were said to be descendants of these sailors.

It is a romantic origin story, and it's not implausible, but in the ab-sence of solid records it's hard to determine if it is real. After all, this is the same island chain in which local lore had it that throwing a cat over your house in a particular direction would ensure a good wind for your sails. Who knows which of the old tales were true and which were con-cocted to explain an anomaly—like the birth of a dark-haired child to blond parents, who told their neighbors that young Angus's Spanish an-cestry was showing?

The day I caught the ferry from Mainland and drove to Pierowall's tranquil half-circle bay, I saw a lot of lovely blue eyes but no dark-haired beauties. In the town archives, outside which lay the massive skeleton of a sperm whale, I read about the Dons and leafed through old photos. Here and there were pictures of olive-skinned youths, looking reason-ably Spanish. But the records didn't reveal much about who they actu-ally were or even if they were native to the island.

Still, behind their eyes, beneath their skin, below the membranes of their cells there is something in the DNA of Westrayans that marks them and no one else. The origin of that distinction is not yet clear, but what-ever it was, the scientific team that detected it in 2012 discovered that even the Orkney Mainlanders don't have it. Is it a legacy of the Dons or something much older and weirder?

It's not just the Westrayans who are different from everyone else. If you examine all the longtime residents of all the Orkney Islands to-gether, they too have something inside their cells that distinguishes them from everyone else in Britain. Throughout the British Isles, in fact, clusters of people carry distinctive traces of ancient events within them. Celtic kingdoms, barbarian invasions, Norse raids from more than a thousand years ago—traces of these distant, almost mythical moments in time are written in the bodies of the good and ordinary people of Devon, Anglesey, Westray, and many other places.

These traces were discovered by an Oxford team that has found a way to read the book of history in human DNA to a level of detail that is completely unprecedented. Indeed, it is the closest thing we have to a time machine. Which is not only to say that it's merely our best shot at traveling through time; in fact, it's quite close to it.

In 1980 Peter Donnelly, an ex-Queenslander who attended Oxford as a Rhodes scholar, was deemed so bright that at twenty-nine he was appointed the youngest full professor in England (and, it's reputed, the youngest full professor in that country in more than a century). Donnelly is now director of the Wellcome Trust Centre for Human Genetics and a professor of statistical science at St. Anne's College, Oxford. Refuting all the stereotypes of outsize genius, he is towering and deep voiced, and if life had led him that way, he could have made an unusually tall but dignified magistrate. Although he trained as a statistician, his work increasingly took him into genetics, and over a period of about ten years he changed from a mathematician who dabbled in genetics to one of the world's leading geneticists.

I met with Donnelly and his colleague Stephen Leslie at one of the genetic world's most pleasantly located meetings, by the beach in Lorne, Australia, some nine thousand miles from Westray. As the afternoon tide hit its low point and turned around again, Donnelly and Leslie walked me though a brief history of genetic research in the twenty-first century, which has so far featured one particularly huge upheaval.

While genes were discovered around the beginning of the twentieth century, it wasn't until 1953 that the double-helix structure of DNA—the stuff that genes are made of—was discovered by James Watson, Francis Crick, and Rosalind Franklin. Almost five decades later a human genome was sequenced for the first time. Despite this enormous and expensive step, the project of linking specific genes to traits or diseases has until recently proceeded painstakingly, one gene at a time. Researchers would pick "their favorite gene," Donnelly said, and investigate only that.

"It wasn't based on the idea that there was only one gene involved in the condition," Donnelly explained; rather, it was simply too expensive to look at anything else.

The problem with candidate gene studies, however, was that a promising result—say, the discovery that a majority of patients with a particular disease appear to share a marker that a group of healthy people do not—might not actually have anything to do with the actual disease. "Now we know that people in Scotland will have genetic variants that differ from people in, say, Tuscany," Leslie explained. "It could be just by chance or it could be by natural selection, but there will be differences between Scots and Tuscans." The danger with candidate gene studies, he said, "was that you thought you were seeing something that was associated with having a particular trait, but actually what you were seeing was something associated with being Scottish or Tuscan." (The other problem with candidate gene studies is that "almost all of those results turned out to be wrong," Donnelly said. "One of the lessons of that era is how bad experts were at picking candidates.")

Around 2007 not only did it become possible to investigate many places in the genome simultaneously, but also the cost of doing so quickly dropped. In a matter of years candidate gene studies were replaced by genomewide association studies. Researchers now had an eagle's-eye view of an individual's entire genome, and they were able to compare tens of thousands of sites in the genomes of tens of thousands of people to identify meaningful correlations with a trait or disease or with the history of a population.

Scientists have known since before they had the technology to measure them that regular genetic differences—what geneticists call "population structure"—probably existed. "For as long as we have measured traits in human populations, we've known that the distribution of those traits vary in different parts of the world, depending on which population is measured," said Donnelly. "For a long time we only knew about a few markers, like blood groups, which we have measured since the 1930s."

Indeed, blood is the classic example: The A blood group is found mostly in Europe, while there's considerably less type A in Asia. The B blood group is more common in Africa than in Europe. The Rh factor, named for the rhesus macaques used to investigate the trait, refers to the presence or absence of a set of red-blood-cell antigens, and it differs too among populations: Rh-negative blood occurs far more often in Europe than in Asia. Even within particular European populations there are differences in blood groups. The Irish Blood Transfusion Service, for example, gets more O-negative when it collects blood in the western part of Ireland than in the east.

Some biological differences between groups may have little to do with how individuals actually live their lives, yet they may still be potent with meaning. They may reveal how long the groups have been separate, how long they have lived in one area, whom they mixed with in the past, and whether their bodies have adapted to local conditions. Combined with historical records, artifacts, or information about the biology of other groups, they may tell us when population differences arose and why they happened. Essentially, one can use the living tissue of human beings to work out what the lives of their ancestors were like up to hundreds and thousands of years ago. It's like William Blake's poem about seeing the world in a grain of sand, except that instead of a metaphor it's real: What you will see is the history of the world in a handful of human cells.

One of the earliest attempts to read deep history in the living body was a project that compared blood types and populations. Historical genetics began by looking at very small parts of the genome, the Y chromosome, which is passed from father to son, and mitochondrial DNA (mtDNA), which is passed down by mothers. (For more about the Y chromosome and mtDNA, see chapter 9.) Methods developed over the last ten years investigate more of the genome and are powerful enough to detect differences between inhabitants of different continents. "I set a project for first-year PhD students," Leslie said, "where I give them a few hundred markers and teach them a statistical method for modeling genetic data and population structure. I give them markers for 120 Africans and 120 northern Europeans. They can write a program in half a day and run it in seconds and work out who's from Africa and who's from Europe just from genetic markers alone."

Yet there is little these procedures can reveal about a group like the pre-twentieth-century population of Britain. "If you use the standard method to try and split Britain, you'll see nothing much. What you'll see is Orkney split, and Wales split, and that's it. You'll see no fine-scale structure at all," Leslie said.

Now, with the advent of genomewide studies, researchers can survey the genomes of thousands of people for population structure. Often this happens in case-control studies, where the idea is to account for ancestral traces in DNA that might otherwise confound medical studies. Donnelly led the 2005 Wellcome Trust Case Control Consortium, a sampling of seventeen thousand genomes that is now regarded as the gold standard for all case-control studies in modern genetics. A year earlier, he and Sir Walter Bodmer, one of Britain's best-known geneticists, had begun another study. Many years before that Bodmer and his wife, scientist Julia Bodmer (who passed away in 2001), had proposed a genetic study to uncover the origins of the British people. Bodmer pursued the idea for years, and when he took it to Donnelly, they conceived of a study that would be important for the investigation of disease in the British population, but the two scientists hoped it would also give them a completely new view on history.

If we consider the entirety of human history, it becomes quite obvious that if people live near one another long enough, their DNA will eventually become blended. In fact, so inclined are people to mix it up with everyone around them that there is always a clear reason for cases when they don't, which is to say that barriers to reproduction must be high. They might be physical factors like mountains, oceans, or extraordinary distances. They might be strongly enforced beliefs. Like the Dons, the Orthodox Jewish community in Brooklyn, New York, as in many other cities around the world, lives in close proximity to other ethnic groups but marry only one another; genetically it's as if they lived on an island.

Still, even when people marry only within their own group or live on an actual island, their DNA is never static. As time passes and DNA is passed from one generation to the next, changes naturally arise in the genome. While some are not passed on, others diffuse through the gene pool. If the group does not mix with others, such changes may become characteristic of that particular group.

In order to have the best shot at finding the characteristic genetic traces of British ancestry, the Oxford team focused on areas with rich archaeology and were selective about the genomes they chose: They looked only at people whose four grandparents were born in rural areas within eighty kilometers of one another. Sampling anyone's genome is essentially the same as taking a smaller sample of their parents' genomes and an even smaller sample of their grandparents' genomes. It was this aspect of the genome in which the team was especially interested.

"Effectively we're looking back in time to what the genetics of that area looked like when those grandparents were born." Leslie explained. "The hope is that if the four grandparents were born in Cornwall, then their parents were born in Cornwall, and so on. We were hoping to get right back to when people didn't move a lot and lived in their own little communities for generation after generation." Many of those who responded to the team's call for subjects were of retirement age, which meant the average birth year of their grandparents was around 1885.

The careful sampling was Bodmer's idea. He began his career in genetics when he studied under R. A. Fisher, a famous founder of two fields of modern science: population genetics and statistics, and he had long been interested in the ways that history shaped populations. It was he more than anyone who believed there was much more to the genetic history of the British Isles than was believed.

The team ended up with more than two thousand genomes, and it fell to Leslie to find a completely new way to comb through them. After applying a method called fine-structure analysis, he took each genome and then compared it, segment by segment, to every other genome in the set. Once he had done this, the genomes were sorted into more than a dozen groups. All the genomes within a particular group were genetically more similar to one another than to any outside the group. No geographic information was used to presort people; the selection criteria were purely genetic.

After Leslie assigned a color to every subject based on his or her DNA group, he placed a pin representing each one on a map of Britain based on the location of his or her grandparents' birthplace. If there was nothing unique in the genetics of each region, Leslie's map of Britain would look like the sprinkles on a cupcake, a random mix of colors. If there were large-scale trends, as the researchers expected, the map would display a messy but suggestive pattern, with perhaps one group of colors clustering toward the east of the country and another skewing toward the west. Leslie hoped to see a refinement of the three or so British groups that had already been identified from other genetic and historic analyses. But when he ran his analysis, he recalled, "I nearly fell over."

The data revealed that there were more than seventeen distinct bursts of color across the map. In some cases a particular group's borders aligned with modern county boundaries or with natural features, like the Tamar Estuary and Bodmin Moor. In most cases the individual groups didn't overlap: Each represented a genetically distinct segment of the population of England in the 1880s. The people pinned to Cornwall, for example, were, stunningly, all the same color—which no one else on the map was. Their color seemed to signify something that was essentially Cornish in their genetics. The same was true for the Anglesey group and for those from Cumbria and Northumberland.

Leslie had been trained as a mathematician, and when he did his doctoral work under Donnelly, he acquired a solid base in genetics, but he was also a mad history buff—widely read in modern and ancient British history. He now runs his own lab at the Murdoch Childrens Research Institute in Melbourne, Australia, but you might still mistake him for a postdoc. Though the tendency of scientists is to cut their ponytails as they move from postdoctoral work to professorship, Leslie's only grew longer over the many months we spoke. His feeling for the complexities of analysis, the weighing of evidence, and the sheer amount of information contained in DNA was always jubilant and contagious, but when he dug into the topic, his focus was absolute.

Leslie remembers the day he first ran the analysis. He didn't see only distinct genetic groups appearing on the screen; the groups represented a set of fine-grained, historical details mapping themselves out of the genetics of Britain. First he saw Orkney break away from the rest of Britain. The order of a group's appearance, Leslie explained, reflected its degree of difference, which meant the Orcadians were the most different from the rest of the British population. Next came the Welsh. Then North and South Wales split apart. Then the south of England broke away from the rest. Then Cornwall appeared as distinct group. "Being able to see that so soon was just astonishing to me," Leslie said. Other groups broke away, like the north of England and Scotland; then came Westray, which turned another color, distinguishing itself even from the rest of the Orkneys.

A colored oval neatly demarcated the exact area where the team thought they might find a trace of the Picts. In the north of Ireland was an unusually mixed group containing two colors. Leslie suspected this was where the English conquered Ulster, sending eighty thousand immigrants to the area from England and Scotland in the seventeenth century to replace the indigenous Catholic Irish. The area became Northern Ireland, and the immigration became known as the Plantation of Ulster. The two groups that were living together but not mixing genetically were the Catholics and the Protestants.

One of Leslie's favorite discoveries showed up next. This cluster lay over the Irish Sea, joining the northeast coast of Ireland with parts of southwestern Scotland. Leslie recognized the digital apparition as soon as he saw it: Modern genetics had mapped the ancient geography of the kingdom of Dalriada, a sixth-century tribal group that spanned Ulster and the Scottish coast. A final group remained. Unlike the other small clusters, this was a massive area of red that covered most of central and southern England. Almost half of the genomes in the study were sorted into it. All those people had something in common. But who were they?

"I just sat there and then I reran it, just in case I'd got it wrong. I ran it over and over again and just went"—Leslie threw up his hands and gave a strangled cry—"Ah, this is amazing! I knew that you could pull apart continents and potentially countries, but to get something at this fine grain, I couldn't believe it was true."

Leslie ran a traditional analysis, known as a principal components analysis, or PCA, on the same data to compare the results to his own. It showed a few of the biggest groups, but after that hardly any differentiation at all.

Leslie showed the new analysis to Peter Donnelly, who recalled, "It was better than my wildest dreams."

I asked Bodmer about the moment of discovery. "It was absolutely staggering that you could get that much differentiation," he said. "I was very surprised, even though I was obviously a person that most expected to find something."

The archaeologist Mark Robinson works in the Oxford University Museum of Natural History, a neo-Gothic mansion on Parks Road. The windows of his second-floor office once stretched twelve feet high, so that the midnineteenth-century astronomer who used to work there could survey the night sky. Now a mezzanine cuts across the room, dividing the windows in half. When I visited him, paper, in one form or another, was piled on every surface, and except for when Robinson offered me a floral teacup with some water in it or produced a head of wheat to illustrate a point about ancient crops, we sat still for hours staring at maps.

When Leslie shared the genetic analysis with Robinson ("I had naively expected that there was going to be a Saxon-dominated group and there was going to be a Celtic-dominated group," Robinson recalled), Robinson began to draw a series of maps of Britain at significant periods of history. We sat before his computer, and he showed me four of them.

The first map was of Britain at the end of the last ice age, between 9,000 and 7,500 years ago, when the first modern humans began to arrive. England wasn't an island then—a huge landmass called Doggerland connected it to the continent. People talk of Doggerland as an ancient land bridge, said Robinson, but it was only a bridge in the sense that Yorkshire is a land bridge between England and Scotland. There were perhaps about 1,100 people living in Britain at the time. Their ancestors had probably taken one of two routes from what we now know as Europe, either walking across Doggerland or traveling by boat across the channel river estuary and up the west coast into Ireland.

Robinson's next map showed Britain between 4000 BC, when agriculture arrived, and the early Bronze Age, around 2500 BC, when Beaker pottery was brought in. By this point Doggerland had been submerged under the sea and people had settled throughout Britain. They built Stonehenge in the south around 2600 BC, while up north they constructed an

even larger henge known as the Ring of Brodgar on the mainland of Orkney. Near Brodgar they built Maeshowe, a hill tomb you can still enter if
you are willing to bend over and walk down a ten-meter-long, one-meter-
high passage. (Neolithic life wasn't all funerary grandeur and mystic ar-
chitecture. A few miles from Brodgar at Skara Brae, you can see hearths,
bedheads, and milk crate–style shelving in a cluster of stone houses that
are joined by an internal corridor, like those of an apartment block. The
houses are five thousand years old, older than the Egyptian pyramids.)
What is known of this latter period has been gleaned mostly from artifacts
and, thanks to the Romans, who invaded in AD 43, some rare written
records. Roman historians described the tribes they encountered, yet
apart from a list of tribal and place names and the few Brittonic words that
survived—*brock* meaning badger, *tor* meaning hill—we don't know a lot
about the British of that period.

Robinson also set up a second screen next to his computer, so he
could put the team's modern genetic map beside the first two maps of
ancient Britain. I could see little correlation between them.

The third map was of Britain between AD 43 and 410. By this period
the Picts were in Scotland, the people we think of as the Irish were in
Ireland, and the Romans had established a presence throughout England extending as far north as Hadrian's Wall and sometimes beyond.
For the most part they were heavily concentrated in the southeast. This
map looked more like the modern genetic map, because the area dominated by the Romans looked a lot like the large red area on Leslie's genetic map. But it was the fourth map that Robinson showed—Britain in
AD 600—that suddenly looked familiar. By this time the Romans had
left, and, as if they hit the light switch on the way out, the written
records went quiet for about two hundred years. I looked from it to the
genetic map, back and forth. They were so alike, I could almost hear
them click.

On the historical map the region representing the area of the major
Anglo-Saxon invasion was shaded one color. On the genetic map the
same region was also a single color. But Robinson pointed at the ancient
fringe kingdoms, Rheged, Elmet, and Dumnonia in the north and west on
the historical map, regions that were home to small Celtic groups that had

hung on to their identity. Rheged, Elmet, and Dumnonia were also clearly delineated on the genetic map: Rheged is in modern Cumbria, Elmet is north-central England, and Dumnonia spans Devon and Cornwall.

"This is the major achievement of the project," Robinson explained. "It gives us a plausible answer, backed up with a lot of data, about what happens at the end of Roman Britain." Which is to say, the team has essentially turned the light back on in the Dark Ages.

Roman Britain lasted for only about four hundred years. Throughout this period, and despite the fact that they had conquered most of the country, few Romans actually lived in Britain. Indeed, many Roman soldiers were actually Gauls who had been conscripted into the army. Although the Roman-ruled southeast was still largely populated by ancient Britons, they had become culturally Roman. Their leaders lived in Roman villas, some of them spoke Latin, their artisans created Roman goods, and unlike their own pre-Roman culture, they had a sophisticated monetary system.

Around AD 410, when the British population was near 2.5 million, the Roman Empire's hold on Britain began to disintegrate when Saxons, Angles, Jutes, and Frisians raided the southeast coast of England. The Romans responded with various tactics, even inviting some barbarians to settle down in the frontier zones. Yet eventually Roman rule collapsed, as it did throughout western Europe, and the empire's leaders withdrew from Britain. "There is an imperial letter," said Robinson, "that says something to the effect that the Britons will just have to try and defend themselves as best they can."

The shock was enormous. Roman Britain had been literate; now, suddenly, no records were made. The local languages began to disappear (modern English retains only twenty-five words from ancient British) and most of the Roman, as well as the pre-Roman, settlement names were replaced with Saxon ones. Agriculture changed completely. The Romans had planted a variety of wheat that was completely replaced by the Saxon version. Other key technologies simply vanished. For hundreds of years the British had produced fine Romanesque pots—lovely, durable glazed containers that held water. But when the Saxons arrived, they brought their own leaky, crumbly containers, and despite the fact that family-sized kilns existed throughout the country, the older, better pot technology was lost.

Over the years many theories have been proposed to explain what happened to the Britons themselves, but, as Robinson observed, "an awful lot of them are complete rubbish," shaped more by the politics of their eras than by actual history. "There was the romantic nineteenth-century/early-twentieth-century view of Saxons coming in with heavy plows, and the Romano-British, who just farmed the light soils of the hilltops, were forced into Wales," he said. "But the Romans had heavy plows and were certainly farming a far larger area than the Saxons.

"You then had the twentieth-century view of complete military conquest by Saxons. The British were supposedly wiped out or fled to Brittany or Wales." Robinson continued.

"Then in the 1970s you have the idea that it's all acculturation and that the freedom-loving Saxons liberated the Romano-British from the imperial system. So they gave up on material culture and things you could buy in towns and switched over to self-sufficiency and the hippie lifestyle.

"Then, around the time of the genocides in former Yugoslavia, and the idea that one ethnic group will slaughter another entered public awareness, you had the theory of Saxon genocide of the Britons."

The prevailing view has been that the catastrophic cultural collapse of post-Roman England is evidence of either a complete massacre or a flight to the west, leaving the population purely Saxon in culture and genes. Now, with the new genetic evidence, one version of what actually happened in the Dark Ages rings much truer than the rest.

The big red group in Leslie's analysis showed a massive, genetically homogeneous area in the southeast of England. History tells us that this DNA would not have been significantly influenced by the DNA of the Roman (genetically Gallic) army. Indeed, the biggest influx of people in the period were Saxons, so it's most likely that the group originated with the Saxon onslaught. This would rule out the theory that the Saxons had no genetic impact on the locals at all.

But how much of the region's modern genetic profile is actually ancient-Briton DNA, and how much is Saxon-marauder DNA? There are many possibilities. If a huge number of Saxons raced in and killed almost all the locals, the proportion might be only 10 percent Briton. If instead only small groups of adventurous Saxons came to Britain, the population could be as much as 90 percent Briton. The only way to know for certain would be to match the DNA against that of the marauders themselves or against a modern group descended from them. In the absence of a handy source of ancient DNA, the team looked to continental Europe for genomes.

"What we decided to do," Leslie explained, "was find a group of European samples and compare the DNA of the British groups to Europeans from different parts of Europe." The team had access to over six thousand samples from a European medical study, on which they ran the same kind of analysis and found fifty-one distinct groups, most of whom had contributed nothing to Britain. (Ancient Italy, for example, was not represented in the modern British gene pool, confirming what historians have already said about the absence of actual Romans in Roman-occupied Britain.) By contrast, the analysis confirmed that Orcadians have a considerable amount of Norwegian DNA, which is predictable from the historical record of the Viking invasions, which began in the ninth century, as well as from other early genetic studies.

As far as the big red group in England's southeast was concerned, the team found that what slowly emerged from the cataclysm of AD 410 was a genome that was 75 percent ancient British and up to 25 percent Saxon, meaning that while the genome of the natives of England's southeast was not completely replaced, it became strongly Saxon flavored during the Dark Ages.

According to Robinson, the likeliest explanation for this genetic state of affairs is that, in the absence of the Romans, the state—and the local population—fell apart, literally. Other historians have suggested that as many as 1.5 million of the 2.5 million Britons died, but Robinson believes that only one quarter million may have survived and that for a long period there was complete chaos. Local warlords fought one another in addition to having to deal with ongoing Saxon raids. People lost crops, suffered from starvation and disease, and had to abandon their settlements. Families could not raise even two children. All the towns were abandoned too, to the point that the Saxons believed they were haunted.

By contrast, the conditions of fifth-century England were just the sort to which Saxon settlers were accustomed. "Saxons were coming in who, although they looked more primitive on the face of it, had societies that worked at the small scale. Every man would have had to bear arms since he was young," Robinson said. But their already being well adapted to a violent, nonmonetary culture does not mean that hundreds of thousands of Saxons flooded in, all in the period around AD 411.

"My view is it only takes about four hundred Saxons arriving each year for a period of seventy-five years and good reproductive success to have them contributing 25 percent of the DNA," Robinson said. "Effectively the Britons still enjoyed a high reproductive success, as their genes were in people who culturally were Saxon." Still, it took a long time for English culture to resume where it had left off. It wasn't until centuries later, when the Normans invaded, that the British once again saw pots that were as good as the ones that had been produced in Roman times.

What of Rheged, Elmet, and Dumnonia? The surviving genetic markers suggest that the ancient fringe kingdoms not only survived Roman rule but also still, to some extent, kept to themselves. We often think of the Celtic population of Britain as a wild, poetic, and singular group of ancient people, though in fact the study confirmed there was not one large group of Celts but many. How did the different groups of ancient fringe DNA survive both Roman rule and the Saxon onslaught? According to Robinson, it was because the Celts in the west had never been fully controlled by the Romans in the first place.

"There was a Roman military presence," Robinson said, "but the ordinary Iron Age peasants were left to get on with things. If they misbehaved, they would be slaughtered. It was like native states in British colonial India. Provided the local maharaja wasn't anti-British, he was allowed to get on administering his society by his own laws." The different Celtic groups were able to repel the onslaught because they still had their own leaders and weapons and were able to organize themselves. They also sustained themselves economically because they still knew how to barter and exchange goods and services. Enough of them survived to have children, who in turn had children, whose descendants today walk the streets of Cumbria and Cornwall.

Robinson already knew from the written record that these kingdoms hadn't disappeared into the post-Roman vacuum, but no one had ever imagined that they might once more be visible—twenty-first-century ghosts shaped by great socioeconomics and good enough DNA.

How did Leslie and his colleagues effectively resurrect people who hadn't been seen for a millennium? The key to their approach was that they didn't go looking for a specific Celtic gene or a Saxon allele. Rather, they looked at *patterns* across the genome, which most analyses ignore. That enabled them to identify very small but significant differences between people who are otherwise overwhelmingly the same. "It's a collection of very slight differences but across lots and lots of bits of the genome," Donnelly said. "You need to integrate all of this information in order to see the whole pattern of subtle difference."

Essentially the genetic groups in the project looked like different blends of very similar material, somewhat like arabica and robusta coffees or like close hues on a color wheel. The Cornish were genetically royal blue while the Devonians were light blue—fundamentally the same, but still categorically different.

What this means is that one of the big stories of ancient British genetics is a tale of people staying put. None of the history would be detectable in the biology if many people hadn't lived in the same place and married someone from their own neighborhood for generation after generation. In the case of Saxon Britain, that means local boys partnering up with local girls from the 400s until at least the 1860s—about fifty-eight generations who married their high-school sweetheart or her peasant equivalent. By contrast, the backstory of the big red genetic group is not so much one of a large homogeneous community as one of an area that lacked significant geographic or historical barriers, a place where DNA has washed freely back and forth since Roman times.

Curiously, although the differences between the different British groups are definitive, they pose no challenge for genomic medicine. "Broadly speaking, Caucasians in the UK are very, very similar genetically," Peter

Donnelly said. "We looked for population structure quite hard early on in disease studies because we didn't know whether we had to worry about it, and it turned out that you don't really need to worry about it. If you are looking at how common genetic variants affect disease susceptibility, geographic differentiation in the UK is not a big problem."

It may be that when scientists begin to consider very rare disorders, the genetic groups will become medically relevant, but in the meantime the story of ancient British genetics is a story about overwhelming sameness and minuscule but definitive difference—all crammed into the same small forensic package.

There were a few anomalies in the analysis that puzzled Leslie. Here and there a single person who belonged to one genetic group seemed to be living in the wrong location with a different group. For example, a genetically Devonian person in the Newcastle region had four grandparents who were all born in the area around Newcastle. If the DNA analysis was correct, it meant that in the mid-1800s, eight Devonians moved to the region and paired up, with each of the four unions bearing children. Of that new generation one child from each of the four couples must have paired up with another, and then these two couples must have had at least one child, who then paired up with the child from the other couple. If the people involved had been Jewish or Catholic, this might have made sense, but there were no obvious religious differences keeping the grandchildren of Devonians from marrying the locals.

Leslie reran the analysis over and over, but no matter how many times he did, the anomaly wouldn't go away. "It was driving me absolutely insane," he recalled, "because I believe the genetics more than I believe anything else. I wanted to understand what's going on, so I started to track the odd individual that really looked out of place."

Leslie went back into the history books and discovered that there was a significant relationship between the two locations in the midnineteenth century. "It turns out that about 150 years ago Devon was a strong mining area (as was its neighbor Cornwall), and this area up in the northeast was a strong mining area as well. There was a miners' strike up north, and the mine owners brought scab labor up from Devon and Cornwall, and these people—there is a huge amount of evidence for this—these people came up and they were ostracized. Because they were strikebreakers, the local people would not talk to them, would not socialize with them, would not marry them. If you go online and look at genealogy Web sites, you can see these family trees of people that have all eight great-grandparents born in Devon, but they are living up there in Newcastle, working in the mines."

Here is where it becomes clear that this kind of fine-grained genetic history is the flip side of the family-history coin. Although genealogy is not widely valued in academia, it meshes perfectly with, and helps explain, social history. These small stories about individual lives reveal the way that individual choices shape the biology and the history of whole populations.

For all the extraordinary answers that the project provides, its greatest contribution may be the number of questions it raises. For instance, despite the power of their terrifying invasions beginning in the late eighth century, the Danish Vikings rapidly disappeared in England, leaving some tantalizing material remains but not a lot in the way of genetics. Culturally they bequeathed only a few place names, typically ones ending in "-thorpe" and "-by," like Coningsby and Cumthorpe, said Robinson. Why did they enter with such thunder but vanish so quickly? How, on the other hand, did the Norwegian Vikings reshape the population of the Orkneys, changing the language, the artifacts, and the genetics?

What about Westray? How did the locals get to be so different even from the other Orkney islanders? As far as the genetic analysis goes, says Leslie, it's unlikely to have been the result of a single shipwreck. And how, after hundreds of years, do the Orcadians remain so different? Even though it's been a long time since the Vikings roared in, there hasn't been enough intermarriage across Orkney to subsume the ancient legacy. What are the social forces that laid down these archaic patterns still reflected in the modern genome? Why haven't the modern Westrayans married the other Orcadians more often? It's not as if they don't have boats.

When I visited Mainland, an Orcadian told me he liked the

Westrayans well enough—it was the people from Wick on the Scottish mainland, "the dirty Wickers," whom you never married. He laughed at the silliness of the idea, the relic of school teasing, and yet . . . We know that human conflicts, beliefs, and borders can structure the genome, but do these minor, trivial prejudices that we don't take seriously date back from further in time and influence our biology more than we know?

What about the Orkneys' DNA? If it's 25 percent Norwegian, where does the balance come from? People have debated the degree to which the Norwegians slaughtered the Picts, Robinson said. But the genetics suggest the Picts live on in Orkney still.

There's also the question of what genetic patterns the invaders brought with them. Not all Vikings came from the same village, nor did all Saxons. If they came from different villages where the locals had married only the locals, their groups may have reflected the different populations that were ancestral to them. There are mysteries upon mysteries here, and with this new method we may now begin to be able to untangle them.

Naturally, this method may be applied in other countries as well. Adding the multidimensional genetic record to the historical and material record may confirm our existing knowledge, as well as contribute completely new insights and resolve old debates. Picking apart the most ancient migrations from the more recent ones will be part of the challenge of the future. In the meantime a general rule of thumb is that the stuff that is everywhere is likely to be the oldest, having had lots of time to spread out.

Robinson is excited about the possibility of taking fine-grain genetic history out of rural areas: "I think if you went into the cities, you would get a great mixture and all sorts of extraordinary things would turn up."

These methods can also be used to learn about any individual's ancestry. "The bits that are documented in my family are Lowland Scots, Welsh, southern Irish, and English," said Robinson. "It's very, very mixed, but I am fascinated to know what it is in my genetic component that resulted in my maternal grandfather having racist abuse shouted at him in the 1920s.

"They shouted, 'Where's your monkey?' because he was a dark

person with curly hair. His hair was white by the time I knew him in the 1960s, but whether he was very dark and Neapolitan in ancestry, or whether he was descended from a Lascar on the coal boats to Cardiff or something, I don't know. His name was Jones. I just have no idea."

There are many ways to read the book of DNA. Leslie and his colleagues' new method complements rather than replaces older ones. Even as the analysis of the whole genome grows ever more sophisticated, scientists and citizen scientists have found increasingly clever ways to wring knowledge from the Y chromosome.

Chapter 9

DNA + Culture

History doesn't repeat itself, but it does rhyme.

—Mark Twain

In 2002 Thomas Robinson, an associate professor of accounting at the University of Miami in Florida, had his DNA tested by an English company called Oxford Ancestors. Robinson knew very little about the origins of his family. A few years earlier an uncle on his mother's side had done some research and traced the family back to Virginia. But Robinson's father had been estranged from his own father, so Robinson knew almost nothing about his background on that side. All he had was a family Bible with some names from his father's family in it.

Robinson began searching records, and after plugging away for a few years he finally got a clue that his father's ancestors had come from the Lake District in England. Wondering if DNA would take the search further, he submitted a sample to Oxford Ancestors, which identified twelve markers on his Y chromosome, offering some suggestions about where his Y might have come from in Europe. Robinson also tested with Family Tree DNA and received similar results.

Sometime after that Robinson got a call from an Oxford Ancestors representative. The caller told him that the company's head scientist, Bryan Sykes, wished to speak to him. "He has some very exciting news he'd like to talk to you about," the representative said.

What the heck is this guy calling about? Robinson wondered. He started to run through worst-case scenarios. The most frightening possibility, he thought, was that he might find out that he was descended from Hitler (Robinson has blond hair and blue eyes). Fortunately this did not turn out to be the case. In fact, the geneticists at Oxford Ancestors had recently compared their DNA analyses with a study that had just been carried out by another lab at Oxford and found that Robinson's Y chromosome, which is passed only from father to son, was Mongolian. This was a surprise to Robinson, but not as big a surprise as the fact that he was related to *the* Mongolian: He was told that he was a direct descendant of the most famous Mongolian in the history of the world, Genghis Khan.

Genghis Khan lived from 1162 to 1227, and although scientists now argue that he was effectively a one-man genomic event, for centuries he was known as "the Destroyer." Khan and his hordes killed about forty million people—the same number as the entire population of present-day Argentina.

In fact the Mongol Empire removed such a large number of people from Asia that it had an unprecedented impact on the planet. Because so many millions of people disappeared, there was an enormous amount of reforestation. The fact that there were more trees meant that the forests absorbed more carbon dioxide, and overall the world's carbon dioxide levels dropped a fraction (0.1 part per million). The Mongol invasion is the only known human event to have had such an effect. (The Black Death killed up to seventy-five million people in the fourteenth century, but because the plague was relatively short-lived compared to the Mongol Empire, which lasted nearly two hundred years, it didn't have the same impact.)

Although Genghis Khan did not remain in the areas he defeated, significant amounts of his DNA did. As he pillaged his way through Asia, he distributed his DNA by raping countless women and fathering many children. When he fathered sons, he passed down his Y chromosome to them. But Genghis Khan didn't just spread his Y by propagating copies of it; he also killed the carriers of other Y chromosomes in those areas (that is to say, he slaughtered all the men), and in so doing he increased the relative presence of his own Y. Because the Mongol Empire lasted through a few generations of Khans, not all the glory goes to Genghis, though. After his death many of his sons ruled sections of Asia, so the Khan Y kept spreading long after Genghis himself was gone.

Almost a thousand years later, in 2003, scientists sampled the DNA of many modern populations throughout the Asian region and found a Y chromosome with a distinct pattern of DNA that was almost entirely exclusive to sixteen groups. They believed the Y they had found must be Genghis Khan's. The geographic distribution of the chromosome suggested that the Y started its spread from Mongolia, where the empire began. The biological pattern of the Y itself suggested that its progenitor must have lived around one thousand years ago, which Genghis Khan did. In addition, the way the Y was distributed in modern-day populations—from the Pacific all the way through Mongolia, on through central Asia, and ending with Uzbekistan—almost perfectly matched the boundaries of the ancient Mongolian empire.

There was one exception to the match between the distribution of the modern-day Y and the demographics of the ancient empire: the Hazara people of Pakistan. According to the geneticists, the Hazaras have the Khan Y, yet they were the only group with that Y to live outside the region that had once been part of the official Mongol Empire. But it may be that the mismatch actually proves the rule, as the team later discovered that the Hazara have an oral-history tradition that asserts that they are descended from Khan himself.

The scientists estimated that the Khan Y is probably carried by sixteen million men today, most of whom live in Asia. When Robinson received his startling call, it looked as if just one, the prodigal accountant, lived in Miami.

Oxford Ancestors found the match between Robinson and Genghis Khan surprising enough that they called him. They told him he matched Genghis Khan's Y on seven out of nine markers. Although markers mutate independently and change over time, there were enough of them in common to suspect it was a good match. How did Genghis Khan's Y make it to the New World? One suggestion was that it had traveled via slaves from Asia to England and from there to the United States.

The company sent Robinson a certificate that read: "This is to certify that Thomas R. Robinson carries a Y chromosome which shows him to be of probable direct descent from Genghis Khan, First Emperor of the Mongols."

They also asked Robinson if he minded if they announced the match publicly. He had no objections.

It wasn't long before the curious connection between the Destroyer and the number cruncher was reported by the *Times* of London, the *New York Times*, the *Miami Herald*, and other newspapers. When the story hit, Robinson was on an Alaskan cruise with his wife, but the press were frantic to contact him. His in-box was bombarded, and his voicemail was full. News agencies flew photographers to find him when his ship called into port. Exciting offers were made, and one TV production company asked to fly him to Mongolia. The Mongolian ambassador to the United States extended an invitation to meet him in Washington. Robinson even got a call from a representative of a chain of Mongolian barbecue restaurants in Texas, who asked him if he would be their spokesperson. Not long after that, Robinson noticed that Oxford Ancestors was offering a new Khan-specific test on its Web site.

As the offers flooded in, some concerned people also got in touch with Robinson. Most of them were genetic genealogists who explained to him that the match had been made on the basis of few markers. In order to confirm the connection, they advised him to get more and different markers checked.

Robinson decided to get a second opinion, so he contacted Bennett Greenspan at Family Tree DNA. The company still had a sample of Robinson's DNA from his earlier test. Greenspan rushed to redo the test and then ran a more powerful test. He determined that Robinson was not a relative of Genghis Khan. In fact, his Y chromosome was not even Mongolian, though it did probably come from central Asia. When he gave Robinson the news, he assured him that his results were private and that he wouldn't tell anyone else about them without Robinson's permission. But Robinson said, "No! You have to tell everyone." So he did.

All the Genghis Khan–related offers faded away, and Robinson happily got on with his life. He now works for a nonprofit company and regularly travels all over the world, although he hasn't yet made it to Mongolia. He later heard that a Chinese journalist complained that he must have had the second Y chromosome test because he didn't want people to think he was Asian. Robinson and his Asian American wife found this especially amusing.

There is something undeniably compelling about the descendants of a historical figure. Consider the actor Anna Chancellor, the eighth great-niece of Jane Austen. It's impossible not to imagine whether she looks like her great-aunt. What about Hitler's nephews, allegedly still alive on Long Island, New York? What is life like for them? Or for Osama bin Laden's children in exile in the United States? In the past decade newspapers have devoted many column inches to the descendants of Charles Darwin, a diverse bunch. One great-great-great-granddaughter, Laura Keynes (also a great-great-niece of economist John Maynard Keynes) converted to Catholicism and is now an apologist for the faith. One of Charles's great-great-grandsons, Chris Darwin, who used to be called "the missing link" at school, gained notoriety for being part of a team that hosted the world's highest dinner party (22,205 feet, on Mount Huascarán in Peru). Now he is an abseiling guide who lives in the Blue Mountains in Australia. In 2009 his sister spent time with a film crew tracing the journey of Darwin's famous ship, the *Beagle*, to mark the 150th anniversary of the publication of *On the Origin of Species*. "Most of the direct descendants have had a pretty busy year," Chris Darwin told a local newspaper at the time. Of the approximately one hundred living descendants of Darwin, one is now an acupuncturist, another is a novelist, one is a botanist who has expertise in the Galapagos tomato, one is an ecologist, one is a dancer, three have received knighthoods, and one works on the TV show *Doctor Who*.

The tale of the accountant and the ancestral Mongol hordes was picked up everywhere in the media because it is, after all, a great story. The tale of the accountant and what turned out to be a non-Mongolian Y is important too. It underlines the fact that bringing together genetics and history involves some particularly tricky science, and because most of us are not scientists, there is some point at which we need to take people at their word—for whatever that is ultimately worth.

It also reminds us that wherever compelling stories appear, commerce will likely be involved. Still, there is much to be learned about our history, and even if some of the most tantalizing possibilities are on the cutting—and less certain—edge, there is a strong scientific foundation for these questions.

It's taken about one hundred years from first understanding the basics of human chromosomes to reach the point at which we're making connections to long-lost ancestors through the centuries. Around the beginning of the twentieth century, scientists discovered that within the bubble of the nucleus that sits within the bubble of a human cell, men and women have twenty-three pairs of chromosomes. One chromosome in each pair comes from the mother and the other chromosome from the father.

Men and women have the same chromosome pairs, with one exception: While both men and women have an X chromosome, women have a second one, but men only have one. In lieu of the second X, men have a Y. This is what makes them men.

Given that only men have a Y, it follows that all men get their Y from their fathers. Recall that Ben Franklin discovered he was the son of a son of a son, going back five generations to Thomas Franklin of Acton, England. Ben would have had the same Y as Thomas. Ben's son William, with whom he was traveling, would have had that Y too.

For a long time scientists believed that the Y chromosome was the same for all men, but toward the end of the twentieth century, they discovered that men's Y chromosomes differ in traceable ways. It became clear that if they looked at a group of Y chromosomes, the ways in which they were different formed a pattern. Most exciting of all, that pattern could be read like a historical record.

What makes the Y unique among all chromosomes is that it is passed down from father to son *as is*. Normally, before they are passed down, the chromosomes in a pair are very lightly shuffled together. A chromosome may swap one or two small segments with its partner, undergoing a process of *recombination*. Unlike all the other chromosomes, the Y doesn't go through the shuffling process, so the chromosome is handed down, again and again, from father to son. The Y's partner chromosome is always an X from the mother, but the Y doesn't mix its DNA with that of the X.

For the purposes of genetic variation and the health of the species, chromosome shuffling is beneficial. It means that when we create a new child, and it gets one copy of each chromosome from a parent, those copies are a good mix of the grandparents' DNA. But the Y has few functional genes and is small relative to the other chromosomes so its lack of recombination doesn't impact variety.

For the purpose of history, the fact that the Y does not recombine is the most perfect happenstance. What it means is that if there are differences between the Y of a father and that of his son, it's not because someone else's DNA got mixed in there too, but rather because something went wrong with the copying process. If we can identify the copying errors and compare them over many different Y chromosomes, we can build a tree of Y chromosomes and see the branching of all men, fathers and sons, starting in the present and going deep into the past, long before we started writing birth certificates or noting children's name in Bibles, long before we even invented writing.

There are two kinds of Y mix-ups that are especially interesting for this sort of research. Sometimes one of the letters of DNA is simply miscopied. This kind of mistake is extremely rare, and scientists who trace such variations are effectively tracking change through thousands and thousands of years. Another copying mistake occurs when a cluster of letters are accidentally repeated. A sequence G-A-T-A might be miscopied as G-A-T-A-G-A-T-A . In most cases, these short tandem repeats, as they are known, don't appear to affect the function of the chromosome, but when you compare the short tandem repeats of related people, they can be used as markers for estimating when a common ancestor lived. (See chapter 14 for an example of a repeat that has a serious consequence.)

If the bearer of a short tandem repeat on his Y chromosome has a son, that child will have the same copies on his Y. At some point further down the generational road, the bearer of the repeat may pass on another miscopied sequence to his son. You can follow the trail of repeats back through time to the original bearer, and they are especially handy for tracking people in the last eight hundred or so years.

There's something fairy tale–like about the way a male lineage has a biological marker. The seventh son of a seventh son is a lucky figure in myth, and for much of Western history the male line had real power in families, where a male heir is always preferred. Curiously enough, mothers have their own special genetic markers too, but they aren't part of the human genome. Rather, they are found in mitochondria, which float in the space between the bubble of a cell nucleus and its outer layer. Mitochondria, typically called the cell's powerhouse, are handy little machines, remnants of a single-celled organism that became permanently entwined in ancient cells long ago and gave rise to most life as we know it—all fungus, plants, and animals. Mitochondria have their own DNA, which is passed down from mothers to their children in the ovum. This means that everyone has the same mtDNA as his or her mother, but only daughters will pass it on. The identity of your mother and her mother and hers before her will always be stamped inside your cells.

When you consider the puzzle of the genome overall, the Y chromosome and mtDNA are almost shockingly informative pieces of DNA. Think of your expanding tree of ancestors, parents to grandparents to great-grandparents, ever doubling as you go back yet one more generation. By the time you've reached ten generations back, there are 2,046 people in your family tree, each of whom (despite believing that the past stops with him or her) has nevertheless contributed to your existence and to your family's genome.

The sheer size of the pool is one reason people often throw up their hands and declare that all those generations of pairings quickly render any single contribution untraceable. But the Y chromosome and mtDNA track back through history in a single line, following fathers' fathers and children's mothers through time immemorial. What else do they tell us?

By the clear waters of the East Branch Pemigewasset River, surrounded by trees of gold, red, and green, there is a cluster of tents with flags of many stripes and colors fluttering in the breeze. It's the Highland Games in New Hampshire, and each tent proudly displays the name of a clan— the MacGregors, the MacDougalls, and the Stewarts. On the tournament field shepherds compete, whistling to bright-eyed border collies that expertly harangue sheep around an obstacle course. A commentator entertains the crowd with tales about the dogs' pedigree. Some of the sires are famous.

Over a whiskey and a big cigar in the autumnal sunshine, I spoke to Donald MacLaren, aka Donald M'Donald V'Duncan V'Lauren, Chief Donald MacLaren of MacLaren and Achleskine, and sporting three golden eagle feathers in his beret to prove it. MacLaren was loquacious and charming, and if you met him and didn't guess straight away that he was a British diplomat before retirement, you would not be surprised to find out. Whenever he is in the clan tent, he is subject to a mild mobbing. MacLarens from all over are excited to speak to him, and there is much joshing about the variety and volume of wine that was consumed the night before and how it might compare to the consumption in the coming evening.

Here in the New World, those of us without documented history, or at least those of us who haven't read the documents that track our history, enjoy something like a definitive sense of closure on the capital-P past. The past of kilts, swords, betrayal, and crowns was then; this is now. In some general sense we are aware that our ancestors lived in the early, middle, and late medieval periods—they must have—but here in the New World the door to all that has shut. Is this clarity or presentism?

The MacLarens were a Picto-Scottish community long before the adoption of a common clan surname. While they took their name from a chief called Labhran who lived in the early 1200s, they trace their ancestry back to King Erc, a fifth-century ruler of Scottish Dalriada, the ancient kingdom that emerged on Stephen Leslie's genetic map. Donald is descended from one of Erc's two sons, King Lorn Mor. The current queen of England is descended from Erc's other son, King Fergus Mor, which means that she and Donald are cousins, although more than fifty generations apart. It is not a relationship he has called on, he said.

Although Donald has no official duties and could ignore the clan, as indeed some Scottish chiefs do, he took it on himself as a young man to go on a tour of North America and visit the far-flung MacLarens. Since retirement he's led MacLaren celebrations many times, and there is much more fun in clan pageantry today than there used to be. For centuries the ancient tribes of Scotland tussled with one another and their overlords for power. Disagreements were often settled with a sword, and terrible betrayals and loss of life were common. A brief trip through the clan history with Donald is like stepping into the George R. R. Martin world of *Game of Thrones*.

From the early twelfth century, the Scottish Crown began to impose feudalism on the clans. Yet despite its attempts to assert authority, there were long periods of lawlessness. "It was a hellish time," explained Donald. "There was complete mayhem in the Highlands. They had 'letters of fire and sword.' If you applied to the Crown for a letter and you received it, it gave you patent to burn your enemies out of their home. There was total economic disintegration; there was homelessness; there were abandonments; there were bandits and reivers and cutthroats."

When the Scottish crown demanded that clans formally produce or apply for legal title to their own lands, Clan Labhran, that is, the MacLarens, refused, and they were reduced to the status of Crown tenants. A 1672 Act of Parliament decreed that clans must formally register their heraldic Arms, but again the MacLarens would not comply, leaving them officially chiefless and landless. Over the centuries, as they engaged in local and national battles, the clan suffered significant losses. The sixteenth century took an especially harsh toll. As the MacLarens attempted to recover from a disastrous battle in which they fought for James IV, a rival clan, the MacGregors, attacked them twice without warning. In 1542, they arrived out of the dark forests at night, murdering twenty-seven men, women, and children. The second attack, sixteen years later, left eighteen complete households burned and murdered. The Mac-Gregors then took over MacLaren land.

It was around this time that the worldwide diaspora of the clans commenced, slowly at first. For some time, Scots had begun to disperse throughout Scotland but now they left for the continent. Two MacLaren families did very well for themselves there: Their founders joined the Swedish army and their Arms were recorded in the Swedish Register of Nobles. In the eighteenth and nineteenth centuries, Scots began to depart for North America, Australia, and New Zealand, many of them fleeing unhappy lives.

It wasn't until 1957 that the ancient chiefly Arms were finally officially recorded. Donald's father assembled a significant amount of evidence to prove that he was in fact descended from the last known

MacLaren chief, and he presented it to the Court of the Lord Lyon in Edinburgh, the Scottish heraldic authority that rules on title and is famously rigorous in its judgment. The court decreed that Donald's father had indeed descended from the last-known chief of Clan MacLaren. When he was made chief, he acquired the legal title to some of the clan lands at Balquhidder that had been lost a few centuries earlier, including the famous Creag an Tuirc, the Clan's rallying point from earliest times. When he died in 1966, his three golden feathers were passed on to his eleven-year-old son, now the twenty-fifth chief since Labhran.

If Santa wore a tartan kilt and were excellent at math, he might look a little like Bob McLaren, the clan's genealogist. A senior scientist in an engineering and manufacturing company until he retired in 2004, McLaren founded the genetic surname project for the clan. At the Highland Games he greets visitors to the tent: an older couple, a rival clan member, a young woman with a tattoo circling her wrist (the words "my darling" in Scots Gaelic). With patient, tireless enthusiasm he tells them about the MacLaren DNA project. There are over 850 members in the project, he said, of whom 754 have had their Y DNA tested. In fact, Bob's project is one of the largest and fastest-growing Y DNA projects in the world.

Y DNA project directors have a lot in common with the eighteenth-century natural historian Carl Linnaeus and his disciples, adventurers who launched expeditions in search of plants and animals, amassing big collections and then carefully creating taxonomies to describe and explain them. McLaren encourages people within the clan to get their DNA analyzed; then he works out how the owners of the Y chromosomes are related to one another by determining what their Y DNA has in common and what makes it different.

McLaren is an extremely skilled amateur. He and directors like him are not only spearheading private science projects but also changing the way the science is done. Over the years they have worked with Family Tree DNA, the first genetic genealogy company to market a Y DNA test, to shape the tests so they include as much useful information as possible.

The company's first offering analyzed twelve segments of Y DNA, counting how many short tandem repeats occurred in each. A test that looked at an additional thirteen segments was developed next. Ultimately both tests gave a pretty low resolution of the past. While lots of people matched one another perfectly on twenty-five segments of the Y, it wasn't clear precisely how. Brothers, fathers, sons, and distant cousins could all have the same result. Not only was the number of segments not useful, McLaren told me, but also there was nothing about the segments that was particularly helpful for tracing bloodlines.

In fact, the first twenty-five segments on the Y test were chosen only because scientists had already developed the tools to isolate them. Geneticists created synthetic primers—enzymes that attach to the flanks of the short tandem repeat—and each different segment required its own primer.

McLaren and his fellow project directors wanted better options, so, as McLaren put it, "We beat on FT DNA to give us more." The company increased the number of short tandem repeats on the test to thirty-seven and they chose more carefully what the additional twelve segments would be, targeting repeats that were known to change more quickly than others. Because they are more likely to change, two Y chromosomes that match in those places may share a more recent common ancestor.

Now, says McLaren, "If you match thirty-seven for thirty-seven with someone, then you're really onto something." People who match on this test certainly have a common ancestor, and they are likely more recent than those who match in fewer places. Yet still, as time passed and greater numbers of people took the test, more of them matched on the thirty-seven-segment test. A few years later a test that looked at sixty-seven places on the Y was developed.

Customers who matched perfectly on the first 37 segments learned that they probably had a common ancestor but not a particularly close one. If they matched on all 67 places, said McLaren, then their ancestor would be a lot closer to both of them. More recently the company has offered an 111 segment test.

Because of Bob McLaren's project, we now know there are many different MacLaren Y chromosomes. This is to be expected, as people could become part of the clan through birth, marriage, or adoption. He showed me a spreadsheet of the clan's Y chromosomes, some of which he believes to be of an ancient Scottish lineage. The point of creating this catalog of Y DNA, McLaren explained, is to help people with their genealogies, not to include or exclude people from the clan.

As for the chief, the documentation that links Donald to his seventeenth-century ancestor forms a compelling body of evidence, but it is just one line of evidence. However, Bob McLaren's analysis seems to agree with the Lyon court's ruling. According to Bob, the chief's Y-DNA results, "place him in a cluster of MacLarens who appear related, but for many of them the relationship is quite a distance back in time, an old Scottish lineage." A fairly large number of people in Bob's project fall into this cluster, and while they have similar values of short tandem repeats, there are enough differences to suggest their common ancestor is from long ago.

"There is no single short tandem repeat that points to the age of Donald's association with the Clan," Bob explained. "Rather, it is the collection of short tandem repeats that lead me to this conclusion. There is one short tandem repeat that is almost unique to this cluster but I need the others to validate this."

I asked Donald about the Y-DNA testing. "It's a wonderful way to complement whatever knowledge we have," he said and then, laughing, added, "If I thought I was maybe a Sinclair or something like that, I might not do it, but I was very happy when the clan genealogist asked me to test."

Is it only clans that share a cluster of Y chromosomes? What about any group of people who happen to have the same name?

Names are powerful symbolic markers. They tell us what has traditionally been most important: Names are unique identifiers, and they encode our ancestry as well. In our surname is our father's name and his before him. Of course, they also have practical applications. In the early twentieth century, when the communist government of Mongolia outlawed surnames, it is alleged that more cases of accidental incest arose, presumably because people didn't know if they were related or not. When the noncommunist government later reinstated last names, the prime minister at the time also claimed that family names reduced crime and increased social responsibility. (Ironically, it was feared at the time that an overwhelming majority of Mongolians who couldn't recover what their actual family name had been might want to adopt Genghis Khan's name.) Certainly it's easier to run a country when you can identify people by last name and family affiliation.

Surnames have the longest history in China, where they began around five thousand years ago. They are about seven hundred years old in England, where they became hereditary in the fifteenth century (and likely earlier). They became hereditary in Scotland a century after that. Surnames are older in Ireland, beginning around nine hundred years ago. In the Netherlands they've only been in use for the last two hundred years. In Turkey they were adopted in the early twentieth century. Iceland is the only European country to retain the older-style patronymic that many other countries abandoned, in which children are surnamed with the father's forename appended to the word for "son" or "daughter," so the surname changes every generation. For example, Axel Stefansson is Stefan's son, but Axel's son Baldur will be known as Baldur Axelsson. In English, names like Johnson, and even just Johns, originated this way.

Many surnames were based on the occupation of their owners, such as Smith, Wright, or Sawyer. It was common for others to be based on nicknames or diminutives, like Redhead. Still others were inspired by local topography or a place name, like York, London, Lake, or Townsend. Some *types* of surnames are especially common to a particular region. In East Anglia, it is said, a number of surnames originated with the local pageant in medieval times. If an actor was well known for his role in the local annual play, such as Herod or the Egyptian pharaoh, he could end up with a surname like King or Farrar.

Even though much water has passed under the cultural bridge since surnames began, their use is still strongly shaped by their origins. For many surnames there's a clear "geographic hearth"; that is to say, where the name originated is still the place where most people carry it. In Great Britain the surnames of Scotland and Wales demarcate distinct regions, and the border created between Scottish and English names is more absolute. Along the geographic Welsh border, the old Welsh surnames are starting to encroach into the English region.

When it became clear that the Y chromosome was passed down the male line, researchers became quite excited about the possibility that a relationship could be established between surnames and the Y. But many were skeptical that names could be treated like genes. After all, names are words and as such are surely affected by the tides of change that sweep through languages and dialects.

But are surnames words in the literal sense? They were originally drawn from language, and the words that gave rise to them (smith, lake, hill, etc.) behave like other words. But once they became names they were, in a sense, arrested, and with their existence as simple nouns over, they became subject to some unusual rules. Now, unlike any other words, surnames can be known by anyone, but they also, in an odd way, *belong* to a select group of people.

Indeed, while dialects and names have a strong connection to geography, they do not have the same relationship. The association has been likened to children's shoe size and reading ability, which appear to be correlated but only because they both correlate strongly with a child's age. One study from the Netherlands revealed that surnames are unaffected by changes in dialects and seem to change according to their own rules, not the rules of language. What about the biological landscape? *Do* surnames track the Y?

In 902 a large Viking band led by a man named Ingimund settled in Wirral in northwest England after being expelled from Dublin. We know that the Viking presence in Wirral was a significant one, because a great deal of Viking jewelry, weapons, and tombstones have been unearthed locally. Many place names in that region are also of Norwegian origin, such as Tranmere (meaning "crane sandbank") and Meols (meaning "sandbank"). Wirral even has a village called Thingwall, the word for a field where Viking men met and conducted their parliament. Neighboring West Lancashire has the same legacy of Norse artifacts and place names, but the story of how Vikings arrived there has been lost in time.

In both areas the invaders clearly affected the culture, but did they re-produce with the locals or did they keep primarily to themselves?

As many of the incoming band were men, the most straightforward place to look for an answer was on the Y chromosome. But even though the Ingimund-led immigration was a large one, there have been other significant immigrations into both areas since, including waves of people from Ireland and other countries. How do you find a specific Y in this population? If the researchers simply took fifty men off the streets of Wirral and West Lancashire and analyzed their Y chromosomes, it would likely tell them nothing—the men or their ancestors might have arrived in the region the previous year or more than a thousand years ago. If there were a signal from Ingimund's day, it would have been drowned by the noise of all the different immigrations since.

Instead of using a random sample of men, the researchers decided to seek out those whose families had lived in the area for a long time. In addition, they sought out men whose surnames existed in the medieval era. Ideally the researchers could have combed through the Viking sa-gas, compiled a list of the invaders' names, and then checked if anyone in Wirral or West Lancashire still carried them. At the time that the sagas were written, however, names were not passed down in families, and recall that English surnames only came into being seven hundred years ago. The best bet for plumbing the social makeup of Viking-era Wirral and West Lancashire was to go back as far as possible and hope that any family groups that had originated a few hundred years earlier might still be detectable when surnames were adopted.

The researchers found lists of surnames in documents dating from the 1300s and the 1500s that many Wirral and West Lancashire locals still carried, like Barker, Beck, Bushell, and Sherlock, as well as less familiar ones like Bilsborrow, Lunt, Tottey, and Crumblehome. For a control group that also had some local heritage, the researchers included men without medieval names but whose paternal grandfathers were born in the region.

The study found that there were indeed different types of Y chromo-somes in both groups, and in the medieval name samples half of the Y chromosomes could be traced to Norse ancestry. Not only did these results reveal that there was mixing between the Vikings and the locals, but they also underscored just how greatly the invaders from more than a thousand years ago influenced the makeup of the local population today. There was a small Norse influence on the Y of the modern control group, and indeed, there's a good chance many of them also had antecedents in the Viking invaders. But by focusing on the men with medieval names, the researchers were able to locate more precisely the descendants of Ingimund and his band and of their contemporaries in West Lancashire.

Another comparison was made with a group of men from nearby Mid Cheshire, which is physically close but does not have the same history of Viking artifacts or Norse-influenced place names. It was found that the Mid Cheshire Y chromosomes were also far less likely to be Scandinavian than those from Wirral or West Lancashire.

In a similar experiment researchers from Trinity College Dublin investigated the Y on the Emerald Isle and discovered that it had little diversity. Why do Irish men look so similar on the Y? One reason, they proposed, was that life in Ireland may have been a bit Genghis Khan–like before St. Patrick arrived. No doubt the Catholic Church has deeply shaped the contemporary culture and population of Ireland, yet a significant part of the nation's genome still carries the stamp of a decidedly non-Catholic fifth-century warlord.

Niall of the Nine Hostages was so named for his habit of kidnapping the kin of other chieftains to ensure their cooperation. He famously fathered many children to consolidate his power, as was the tradition of the day. In 2005 researchers at Trinity College Dublin investigated a Y that was shared by 17 percent of Irish men in the country's northwest. Because the geography of the modern Y overlapped with the region once ruled by Niall, the researchers believed the Y might be his legacy. They narrowed the subject population further by testing fifty-nine men whose surnames originated in Niall's clan (including Donnelly, O'Donnell, O'Gallagher, O'Doherty, Flynn, Egan, and O'Rourke) and found that the putative Niall Y was particularly frequent in this group. The Y also dates back to more than 1,700 years ago, which puts it in the Niall era.

As with the Genghis Khan Y, the frequency of the Niall Y was amplified by the social mores of the period. Divorce and polygyny were

socially acceptable for hundreds of years in Ireland, even after it became Christian, which enabled Niall's male descendants to spread his Y
far and wide. In one famous example, Lord Turlough O'Donnell, a descendant of Niall who lived at the turn of the fifteenth century, had eighteen sons with ten different women. Those sons, in turn, went on to have
fifty-nine sons of their own. Because the Irish diaspora of the nineteenth
century was so large, and because the Niall Y was already so frequent in
Ireland, it's thought that two to three million men worldwide can trace
their Y to Niall.

Researchers began to use surnames as an index of relatedness in the
late nineteenth century. In the years since they have been utilized, with
some success, to estimate how much inbreeding has occurred within a
population. In the 1970s a doctor in Montreal established a connection
between Y chromosomes and surnames when he tracked a novel mutation on the Y of a modern French Canadian man and his extended family back through generations to a pioneer of the same name. It took until
the beginning of the twenty-first century to show that clusters of repeated
segments could be tracked through time, along with surnames. In one of
the earliest Y/surname experiments that tracked short repeats, researchers in 2000 found that a particular kind of Y was common among Irish
men with Irish names but much less common in Irish males whose
names were obviously English or otherwise non-Irish.

Since the basic link between the male chromosome and the male-
line surname was established, it's become clear that because history
shapes names, and because history and biology affect the occurrence of
Y chromosomes, there is no single straightforward relationship between
them but rather many subtle and variable patterns. For example, if you
want to investigate the old surname/Y connections, uninterrupted by migration in the modern era, you are better off looking in rural communities where people do not often move into the community. But that doesn't
mean you can't learn about the character of a city from its names too.
The centuries-old, worldwide immigration into London is underscored
by the fact that it overwhelmingly has the most unique surnames of any
place in Britain.

Most curious is the way that Y/surname patterns differ between countries. In Britain, on average, a man who has the same surname as another is significantly more likely to have a similar Y chromosome, and therefore a common ancestor, than he would with someone of a different surname. But there's a twist: The Y similarity depends on the frequency of the surname within the population. If you are a Smith, for example, the rule does not apply.

The name Smith was adopted many times over in England—all, of course, based on the occupation of the bearer. A man called Smith is no more likely to share a Y chromosome with another Smith than he is with anyone else not called Smith. While Smith is, unsurprisingly, the most common name in England, any English surname that is held by at least ten thousand people is effectively a Smith-type name. (This includes the Kings, the Brays, and the Steads, for example.) No doubt, if surnames were just coming into general use now, Smith would be one of the rarer names, and we would perhaps be encountering more John Analysts, Jack Realtors, and Susan Hackers.

Setting the Smiths aside (all 600,000 of them), there are many other names that are less common, and in fact, a large pool of them are distinctly unique. There may be all sorts of biological and cultural reasons why they are rare, but one of them is simply that the source name was unique to begin with. Another possibility is that the male line was simply not successful. Instead of having ten sons, like his medieval neighbor John Johnson, Bill Billson might have had only one. If Billson junior grew up to father ten daughters but no sons, that would mark the end of that Y and the good Billson name.

For the English names that have survived with very few holders, there is a strong link between the name and the bearer's Y chromosome. In fact, the less common the surname is, the more likely it is that any two men with that name have a similar Y, as is the case within name groups like Werrett, Titchmarsh, and Attenborough. The Attenborough men are the most homogeneous of them all; 87 percent of them have descended from a single man and share the same Y chromosome, which is found in only 1 percent of the rest of the population.

So strong is the link between the surname and the Y in such cases that it's been suggested that, for many of England's unsolved cases of murder and rape in which a Y DNA sample has been taken from the scene but no suspect has been identified, the Y profile could be matched against a database of Y + surname matches, generating a list of possible names for the bearer of that Y and significantly narrowing the pool of suspects.

In Ireland, by contrast, a man is thirty times more likely to share the same short repeats on his Y with another man of the same name than with a man of a different name, regardless of how popular that surname is. The surname Ryan, for instance, is held by approximately 1 percent of the Irish population (Smith is held by 1.3 percent in England), yet there's a forty-seven-times-greater chance that any two men who are called Ryan will have the same Y than any two men who are chosen at random.

Even though there are sixty thousand Irish Ryans, researchers believe that many of them have a Y in common because when surnames began, there were only a few, and perhaps even only one, Ryan. This could mean that most of the many Ryans of modern-day Ireland descend from this founder. The O'Sullivans, O'Neills, Byrnes, and Kennedys all appear to have had a dominant founding Y as well.

The Kellys and the Murphys are more like the Smiths. While two random Kelly men are 4.5 times more likely to share a related Y than they would with a non-Kelly, there are many clusters of men with different Y chromosomes within the Kelly name, as there are among the Murphys. It's likely those surnames were more ubiquitous early in the country's history. The McEvoys, for their part, apparently had two dominant founding Y chromosomes, a theory that is supported by records revealing that when the name was anglicized, two ancient families, the Mac Fhiodhbhuidhes and the Mac an Bheathas, were drawn in under the same banner and both became McEvoys. History also indicates that fully three Irish surnames—McGuiness, Neeson, and McCreesh—are all anglicizations of the same Gaelic name Mac Aonghusa (son of Angus), which DNA evidence confirms, as all three groups overlap strongly on one Y.

Surname patterns in England and Ireland emerged in the landscape where surnames were invented, but in the United States only Native American names have not been imported. Nevertheless, patterns of immigration and neighborhood formation in the New World can be illuminated by what we know about the origins of the surnames of the people who settled there. In the absence of other information, you could make a good guess about the antecedents of much of the modern populations of the United States, Australia, and Canada from their single most common surname—which, again, is Smith.

At the Highland Games I met Glynis McHargue Patterson, a MacLaren by marriage and an administrator of the McHarg/McHargue YDNA Project.

A few years ago she was contacted by a man called McHarge who lived in Canada. He had been born in Knoxville, Tennessee, and he knew his grandfather had lived there but he couldn't discover anything about his family beyond this. Glynis found the man's grandfather using census records and traced him back through the male line to an Anne McHarge, who lived in a small town in the mountains of Tennessee in 1850. But the trail ended there; she could find no trace of Anne's husband. Where was the missing Mr. McHarge?

With more research Glynis learned that there was no missing male McHarge: Anne never married, so McHarge was her maiden name. Glynis suggested that the Canadian McHarge take a Y-chromosome test. He would not have a match in the McHarg/McHargue YDNA Project, because, of course, Anne didn't have a Y chromosome to pass down. Still, he might get some more clues about where his paternal line came from.

The man took the test, and Glynis was right—he did not match the McHarg/McHargues. Curiously, though, he came up with two sixty-six of sixty-seven marker matches to another family tree DNA surname project for the name Bieble. Glynis went back to the records and one day realized that the family who lived two doors down from Anne McHarge were called the Biebles, and they had a seventeen-year-old-son. She encouraged the Canadian to get in touch with the Bieble surname project administrator and ask him about it.

It wasn't long before the two men had come up with a likely explanation. Indeed, the Canadian's Y was most definitely a Bieble Y, and the administrator, who knew a great deal about the family history, confirmed that the reputation of his ancestor, the seventeen-year-old neighbor, had lasted as long as his Y. Anne McHarge's neighbor, the Canadian told Glynis, was known as something of a ladies' man.

Confirming the Bieble connection meant the Canadian not only learned about the male side of his family but was also, because he connected with the Bieble project, able to look back much further in his genealogy. He was beside himself, said Glynis, who added that genealogical detective work consists of "genealogy, genetics, and . . . pure dumb luck."

Since I started spending time with detectives of the personal past, I'd heard quite a few stories about the role historical neighbors sometimes play in cold-case illegitimacies (which genealogists politely refer to as "nonpaternity events"). How often did such births happen? For a long time the estimate was 10 percent in every generation—a figure that has been used to, among other things, discredit all of genealogy. Based on his own Y research, Bob McLaren believed that 10 percent was too high, and since academic researchers began to investigate Y and surname patterns, their findings have backed him up. A number of different studies have converged on a number of well under 5 percent and much closer to 1 percent. The long-term effects of even that low a figure could still be significant, but it's hardly enough to discredit all documentation.

The patterns created by surnames and Y chromosomes are still being investigated, and the role that different social and biological forces play must be untangled. So far it seems that even if a surname group overwhelmingly shares one kind of Y, it doesn't necessarily mean that the Y belonged to the first male to take the name. It's possible that other Y chromosomes were introduced into the family, perhaps even hundreds of years after the surname was founded, and were more successfully reproduced over the generations until they came to dominate. For example, there's only a 10 percent chance that the dominant Y in the Attenborough group is the Y of the original Mr. Attenborough. It's more likely to have come from a more recent and rather successful inductee.

In Ireland researchers found that for all the surnames they examined, even when there was one dominant Y, it was only ever held by half of the modern subjects, at best. That means that all the surname groups are groupings of different Y chromosomes. The variable topography within the modern Y/surname groups is a testament to the dynamic forces that have shaped people's lives over time. In addition, adoptions, illegitimacies, and other unique interactions affected the journey of the name and the Y through time. And that's just the Y.

Once you start following the pattern made by one segment of the genome through time, you realize that even though chromosomes are transmitted as a package, twenty-three from each parent, the smaller segments of DNA that a chromosome is composed of can travel quite separately from one another. Mitochondrial DNA is like the Y, because a mother's mtDNA is not recombined with a father's, so the father's is not passed down and the mother's goes to all her children. If she has lots of daughters, it spreads out from there.

The journey of the X chromosome is unique too. If a woman has a son who has a daughter, then her X does a funny kind of hop through the generations. My sons get an X chromosome from me, which is a mix of the X chromosomes I got from both my parents. If one of them has a daughter and passes his X on to her, it won't be reshuffled. Remember, his X is paired with his Y, and the Y does not recombine, so for this one generation, the X behaves a little like the Y.

My future granddaughter—let's call her Trixie—will have an X from her father and an X from her mother. Here's where the skip-hop comes into it: The X from Trixie's mother is a mix of Trixie's mother's parents, but the X from Trixie's father is a mix of his mother's parents, making Trixie ever so slightly more like some people who are three generations back on her father's side than she is like the same people on her mother's side. Or, to put it another way, Trixie's paternal grandfather contributes nothing to her X.

The separate lives led by these different segments of DNA give us a glimpse of the fine-grained social processes and physical interactions that together add up to form the genomic history of the world. They force us to contemplate how alike we are *and* how unique we are. Above all they underscore how brief our personal moment in that history is. I once read that the flow of genes through time is like a great river, and individual lives are just eddies in the stream. When an eddy forms, the current is paused for a microsecond, and there we are—an assemblage of many different bits from many different sources—and then the stream pours on, and all those bits go forward in time, except that we no longer travel with them.

Apart from the Y and X chromosomes, the rest of the genome is called the autosome. These twenty-two chromosome pairs are the ones that get recombined before they are passed on. As scientists have developed different ways of analyzing the autosome, more and more ordinary people are beginning to access the secrets inside their own autosome, as well as larger truths about the human network.

Chapter 10

Chunks of DNA

Study the past if you would define the future.

—Confucius

On a winter's night in Utah in 1999, Scott Woodward's phone rang at 2:00 a.m. The father of four teenage sons, Woodward leaped to answer it and heard an older man at the other end ask, "Is this Scott Woodward?" Then, "Do you know anything about DNA?" Woodward was, in fact, a professor of molecular genetics at Brigham Young University. The caller said his name was James LeVoy Sorenson. He was from Utah but had traveled to Norway in search of his ancestral origins; he was trying to find a connection to the past, but he wasn't getting very far. Sorenson wanted to know whether Woodward would be able to analyze the DNA of every individual in Norway for him. And if he could, how much would it cost?

Woodward, a pretty laid-back guy, said he'd think about it and he agreed to call Sorenson in a few weeks. He went back to his lab, and he hunted down some people who had worked for Sorenson. "Is this guy for real?" he asked. "Well, he's been known to do crazier things," was the reply. Indeed Sorenson was the richest man in Utah and one of the richest men in the world. He'd founded thirty-two different companies, personally held sixty patents, and in his lifetime had made well over a billion dollars. The consensus seemed to be that Sorenson's projects tended to involve big ideas that cost a lot of money—and sometimes made a lot of money in return. One of these ideas was the disposable surgical mask, which Sorenson invented in the 1950s.

So Woodward thought about it. How much *would* it cost to analyze the DNA of everyone in Norway? How many people lived in the country?

How could he obtain a blood sample from each one of them, and how much actual blood would that be? These were compelling questions, but Woodward wasn't entirely comfortable with them. It was one thing to spend millions of dollars analyzing the DNA of an entire country, but it was another to determine if it was a worthwhile exercise. Would taking all that time and employing all those people be the best use of Sorenson's money? Did the project have merit as well as ambition?

Of course, all professional scientists have to be budget conscious, as there's only so much research funding to go around, and in order to get any of it, they have to show they can do more with it than the next lab jockey. But even if scientists weren't competing for financial resources, they would—by nature—be cautious with them. The essence of all successful research involves wringing the most knowledge from the fewest experiments, and Woodward wasn't clear about how much knowledge was enough to justify an undertaking of the scope of Sorenson's proposal.

At the same time that he began putting together an estimate for the Norway project, Woodward started working on a second idea to propose to Sorenson. Woodward's study would cost the same amount of money but would give Sorenson—and the world—a lot more genetic bang for their eccentric-billionaire buck. Still, he assumed that neither project was likely to get off the ground, as the money that would be necessary was, as far as genomic analysis went, a completely unprecedented amount.

Two weeks later Woodward sat at Sorenson's desk, and Sorenson asked how much it was going to cost him to genotype Norway. "Well, I can tell you what it will cost you," said Woodward. "But you can't afford it."

"Oh, really?" said Sorenson. "How much would that be?"

"It will cost you half a billion dollars."

Sorenson shrugged. "I can do that."

Whoa! Woodward thought. *I have set my sights way too low.*

Around the time that Sorenson called Woodward, the wider world was just becoming aware of the extraordinary potential of DNA. For some years scientists had been working out how to trace the movements of long-dead people through the Y chromosome, and it was becoming clearer that if you knew how to read it properly, the DNA of even a random set of individuals could open up tracts of human history that had been entirely obscure. For exceptionally savvy observers like Sorenson, the letters of DNA must have looked like the proverbial bread crumbs, scattered by individuals in the past as they wended their way around the planet.

What was even more striking about this new way of looking into the past was the fact that individuals had a fundamental role to play in it. Genetic history didn't just survey the broad strokes of history and the lives of a few notable individuals—it built the picture up one human being at a time. This meant that not only could you see far back into the history of the world, but you could also see where your DNA came from and, indeed, traces of who you came from in it.

Where Sorenson's project was concerned, Woodward explained that analyzing the genome of everyone in Norway would only reveal so much. It would tell a lot about the history of Norway, and it would tell Sorenson where he fit into it. But for the same half billion dollars, Woodward said, they could answer not only Sorenson's personal question but also many, many more at the same time.

Woodward proposed instead that they analyze the genome of two hundred individuals from each of five hundred different populations around the world. That collection of 100,000 genomes would form a microcosm of the human race. The DNA would be representative not just of everyone alive today but also of a massive fraction of everyone who had been alive, particularly in the last few hundred years. The goal, Woodward explained, was not just to analyze and match up the DNA of that many people but to get at least four generations of family history from each person as well. They were going to use science to personalize history. But first they had to ask 100,000 people for some blood.

For not entirely coincidental reasons, it turned out that Utah, with its massive holdings of genealogical data, was the best place in the world for anyone to have this kind of idea. The team started with the "lowlying" fruit at Brigham Young University, explained Woodward, who put some of his students onto the project. "Everybody that is currently in Utah was not in Utah 160 years ago, so their ancestors weren't here, other than the Native Americans. So when you combine genealogical information with the DNA information you get out into lots of different places very rapidly." In the first couple of weeks they had collected three thousand samples of blood, and within a year they had collected ten thousand, each with two hundred to three hundred years of genealogical and geographic information attached to it.

When I visited Woodward's lab in Salt Lake City twelve years after his call from Sorenson, I asked him how he had taken the project beyond the small city near the Great Salt Lake and convinced people to participate. Once they had exhausted Utah's diversity, he said, they had capitalized on the Family History Centers of the LDS, which were all over the world. They also approached genealogical societies.

Woodward's team asked people to donate their DNA as an act of philanthropy. While the study was independent of the LDS, it must have helped that the church already had well-established patterns of charity based around family lineage. Indeed, James LeVoy Sorenson's objective in traveling to Norway in the first place had been to find ancestors to whom he could offer church membership.

As large as the Family History Center network is, it does not contain data for the entire world, so once Woodward's team exhausted the church network, they had to find populations who weren't just willing and interested but whose history was telling in some way. They went to Africa and collected 10,000 samples. They went to Asia and Kyrgyzstan and many other countries. One of their earliest trips was to Mongolia, a major crossroads for the Silk Road, where they collected three thousand blood samples.

Woodward's organization, the Sorenson Molecular Genealogy Foundation, was one of the first genetic genealogy companies. It eventually acquired more than 100,000 samples of DNA from all over the world. SMGF analyzed Y chromosomes and mtDNA, which for that era was an incredibly expansive survey. Still, such is the speed of change in this field that what they accomplished looks limited in comparison to what we have today. SMGF was sold in 2012 to Ancestry.com, the biggest

genealogy company in the world. Since its acquisition, the original project has grown even beyond the epic vision of the Mormon billionaire.
Scott Woodward now heads up a group that links millions of documented
lineages with more than 700,000 spots on the genome, including autosomal DNA along with Y DNA and mtDNA.

The amount of time between the discovery of genetic and genomic
knowledge and the transfer of that knowledge to the public has been extraordinarily brief and possibly completely unprecedented. As of 2014 a
small handful of well-known companies—Family Tree DNA, 23andMe,
and AncestryDNA.com, as well as National Geographic's Genographic
Project—and services offer a selection of DNA tests and genealogical
connections to the general public. Depending on which service you buy,
you can have up to 111 segments of your Y chromosome analyzed (if
you're male), some or all of your mtDNA, and most of your twenty-two
non-sex chromosomes. (23andMe also looks at health and traits; see
chapter 14.) Once you have your data, you can also use do-it-yourself or
noncommerical sites like Promethease, SNPedia, Interpretome, and Do-
decad to interpret it. The genetic genealogy companies often investigate
many more places in the genome than were considered in academic studies of just a few years ago. When it comes to personalized genetic research, many individuals can afford what most researchers can't, which
means that the banks of genomes held by Family Tree DNA and other
companies are essentially crowdfunded, uniquely valuable libraries. Indeed, the biggest collections of Y chromosomes and mtDNA in the world
are found at Family Tree DNA, not at research facilities.

What do these extraordinary companies look like? After all, the genome is a treasure house; it's the Library of Congress many times over,
all stacked on top of the long-lost library of Alexandria. I wanted to see
the genome made visible, so I visited Family Tree DNA, the first genetic
genealogy company, founded in 1999 in Houston, Texas. On a day of
crushing heat I drove from Houston's airport through the urban sprawl to
an undistinguished office building to meet with Bennett Greenspan, a
showman with a Texas accent and a nontrivial moustache.

Greenspan settled in for a talk, but when I asked him if I could see where they analyzed the DNA, he said, "It's pretty boring." "No, no," I insisted, "I've never seen a DNA lab before. It will be fascinating." He took me to the suite's back rooms, where samples of DNA sent in by customers were opened. The room looked exactly like the back room of an office complex in exurban Houston. It was small. There was furniture. There were windows too. In one corner a machine made slight chugging noises. Greenspan was right. Once you have the right machines, you don't need a lot of space or gleaming paraphernalia to uncover the mysteries of the human race.

Greenspan showed me where the sample, usually provided by the customer in the form of saliva or a cheek swab, first goes into a machine that isolates the DNA from everything else. The next machine was what he called the shake-and-bake or, more formally, the PCR (for polymerase chain reaction) machine. The DNA goes in, he explained, and then they "heat it up and cool it down, and heat it up and cool it down, and every time they do, it makes more and more DNA. The idea is to magnify the amount of DNA so you end up with more needles than haystack." Once that is done, the DNA is placed in "the world's most expensive freezer," which holds about eighty thousand samples frozen at minus twenty degrees Fahrenheit. When the samples are taken out to analyze, it's done by robot to eliminate potential human error. Next comes the analysis. Depending on the test, the sample goes through yet another machine, this one with fluorescent magnetic beads. If the segment of DNA that they are looking for is present in the person's sample, it will attach to the beads.

Typically companies provide a graphic representation of a customer's chromosomes, and if members of the same family have their genome analyzed, they can superimpose the respective images over each other and see exactly where their DNA overlaps. It may well be obvious to you that the shape of your sister's eyes is the same as yours, but now you can also see that you share an identical segment of chromosome 3, among others, and that you are literally constituted from the same raw materials at many places all over your genome.

All the companies offer an interpretation of what your genome reveals about your ancestry. You might find, for example, that your Y chromosome is usually found only in men from Africa or Europe or India or Mongolia. You may discover that a certain sequence of letters in your autosomal DNA is typically found in someone with Finnish heritage or Korean ancestry. Only a few years ago the world of science was turned upside down when it was discovered that in ancient times two nonhuman species contributed to the human genome. Now, for a small fee, some companies will analyze how much of your genome comes from Neanderthals. (More about Neanderthals and our other sister species, Denisovans, in chapter 12.)

Greenspan gets personally involved in helping customers solve mysteries from the past. One woman wrote to him and said that even though her family had no Jewish ancestry that she knew of, her grandfather's given name was Herschel, and her father and his brother were given Herschel as a middle name. A few years earlier she had found an art print hidden behind a painting that had belonged to her grandfather. It was dated 1891, and someone had written "L'Shana Tova," a Hebrew new year's greeting, on it. Family Tree DNA tests of her DNA showed that it was likely that she was, in fact, Jewish.

Another woman who was raised Catholic in Spain came from a family with an oral tradition of Jewish ancestry that dated back twenty generations. Greenspan found that the woman's mtDNA matched people who were from places like Spain, Greece, Algeria, Bulgaria, and Turkey and that they were all, indeed, Sephardic Jews. She asked Greenspan to write her a letter that she could take to a Jewish court to apply for issuance of a letter of return, a formal acknowledgement that she was Jewish.

"I would have discounted her story pretty much out of hand as almost unprovable or almost unbelievable that a family could for twenty generations pass on that they were part of this minority group," Greenspan told me. "I always worry when we deal with oral history without dealing with the DNA because I think that the difference between telling stories and telling truth is multiple sets of evidence."

After fewer than fifteen years in business, personal genetics companies have made possible a genetic-based social networking that makes the social networks of the early 2010s look like child's play. Customers of

23andMe, Family Tree DNA, and AncestryDNA can discover an extraordinary collection of genetic cousins and track not just their families
but specific segments of DNA in their genome. By combining the records research of millions of people with the analysis of their DNA, users can build an actual copy of their human network over the last few
hundred years or longer and find their individual place within it.

Companies alert customers when another customer has one or more
largeish segments of DNA that look exactly the same. Geneticists say
these shared segments have "identity by descent," meaning they are the
same because we got them from the same person. The implication is that
the segment has been inherited by both customers from a common ancestor. Accordingly, the company will describe the DNA matches as a
"relative" or a "cousin." Fundamentally, the more closely related you are
to someone, the more segments you will have in common, and the longer
they will be.

It is said that autosomal DNA can take you back at least five generations. The probability of identifying a third cousin using autosomal DNA
is roughly 90 percent, a fourth cousin 50 percent, and a fifth cousin 10
percent. Greenspan said he knows of people who have definitively identified eighth cousins with autosomal DNA. It's hard to believe that these
boundaries won't be extended further as the science develops.

With a large enough group of descendants in a single family line, it
is even possible to rebuild the genome of the dead. The genome of each
living person could be used as a virtual puzzle piece with which to reas-
semble much of the genome of the group's common ancestor. It is theoretically possible, though no one has yet done it, to reconstruct
individuals who have left no trace in written history at all. "With enough
people," Woodward said, "you could project back and rebuild the population of Leigh in Lancashire, England, in 1850."

A few years ago Blaine Bettinger, an intellectual-property lawyer
with a background in molecular biology, began a personal search for the
DNA of his great-grandmother. Born in 1889, she lived so long that Bettinger, who is thirty-eight, met her as child. She was adopted, he said, so
he knew nothing about her genetic background. Yet she was "an incredibly strong woman who had a big impact." There are traits in Bettinger's

family—he thinks of them as ripples—that seem to emanate from her. In order to explore her and her influence on his family, Bettinger started collecting the DNA of her descendants so that he could isolate some of what they inherited from her.

With the help of two of her grandchildren, he has been able to identify thirty-five segments of DNA spread over almost every chromosome that came from her *and* her husband. His next step is to find relatives who match through those isolated segments so that he can prize apart which segment came from her and which came from him. As well as convincing members of his family to participate, Bettinger uses cousin matching to find others who overlap on those segments.

Wading through the personal-genomics options and analyses can be daunting for people who have not thought much about their own genes before. Bettinger ordered his first DNA test in 2003, when companies offered to read around 175 markers on the autosome; now the tests examine just under 1,000,000 markers. As a result, Bettinger has become a leader in the genetic genealogist community, part of a select group of individuals who help people understand their cousin networks and what their DNA may tell them, much of it through his popular blog, TheGeneticGenealogist.com.

CeCe Moore, another genetic genealogist and blogger (YourGeneticGenealogist.com), became interested in the subject when she began to put together a family tree for a niece who was getting married. "Little did I know, it's addicting," she said. "And I'm one of those obsessive-compulsive types." Moore used to work as a TV producer but is now a genetic genealogy consultant for the television shows *Finding Your Roots*, with Henry Louis Gates Jr. and *Genealogy Roadshow*. She has essentially developed an entirely new class of career, not just explaining and interpreting the ins and outs of DNA but being a genetic detective who helps clients find the missing. Now she eats, lives, and breathes DNA all day, she said, and often through the night as well.

One of the growing uses of genetic genealogy is for adoptees to find families, and Moore, who is the administrator of the Adopted DNA Project at Family Tree DNA, is regularly asked for help. "I get e-mails from people literally every day who found out they're not who they thought they were genetically," she said. "They come out as a half-sibling to their sibling, who they thought was their full sibling. This is happening all the time. It usually turns out that the father isn't the father. I've also been contacted about what looks like a baby switch in the hospital." One of Moore's earlier cases involved tracking the family of someone who was left on a doorstep in 1916, but in the past year she has been contacted on behalf of half a dozen individuals who were abandoned as newborns, some found in dumpsters.

Sometimes people don't get the ancestral result that they expect. They assume that they are Irish, but the test says their DNA is more like that of a Russian Jew. "This is just my experience," Moore added. "It could be that people are drawn to testing who always felt like they didn't fit in or always had a question in their mind, but the numbers are very high."

By starting with the segments of DNA that people have in common on one or more chromosomes in genetic genealogy databases, Moore has found missing siblings and even parents. First she tracks back through the family histories of the genetic cousins to find a common ancestor among them, then she will attempt to work her way down again from the common ancestor to find the individual's parents. "You build the tree up and then you build the tree down," she said. "If a predicted second cousin shares great-grandparents with a person looking for their birth family (maybe great-great-grandparents, if they just happen to share a little more DNA than expected), we can build that tree down; we can see who lived in the right place at the right time, who was the right gender, and you can sometimes solve it." Sometimes Moore has found a direct match, where it is quite obvious from the amount of DNA that two people have in common across many chromosomes that they are siblings or parent and child. She recommends that all her clients send their samples to 23andMe, Family Tree DNA, and AncestryDNA, which together have databases of over one million autosomes. "You have to fish in all three ponds." Sometimes all that can be found about someone is her general ancestry: A recent client discovered that she was Mexican, which she had no idea was the case.

While many individual mysteries have been solved with genetic genealogy, if we change the focus to entire populations, shared chunks of autosomal DNA can take us much further back in time than even eight generations ago.

Modern culture is good at retaining general knowledge from the recent past, and since the invention of archaeology and paleoanthropology we've been able to recover some of the distant past too. Still, much of what we know about the ancients had to be effortfully uncovered. Though we may feel a kinship with figures from the nineteenth and eighteenth centuries or even the Picts, who lived a millennium ago, or the Romans from a thousand years before that, we have lost almost all sense of connection with most people who lived more than two thousand years ago. We have a reasonably good idea about some of their knowledge, which we've been able to work out from the traces they left. But we have little sense that that knowledge was passed on to us, not in the same way that we can confidently recognize the innovations and legacies of various significant characters over the last five hundred years, from Leonardo da Vinci to the Wright brothers.

Yet we are now on the cusp of boom times in deciphering the deep history of the world, because in addition to the commercial genetic genealogy companies, many scientific teams are devising different methods to get even further back into the past. They promise not only to illuminate chronologically and emotionally distant times but also to clarify the ways in which we remain connected to them. At the same time that the British team was devising a unique method for revealing the regional genetic blends of Britain, Peter Ralph, a professor of biology at the University of Southern California, and his colleague Graham Coop of the University of California at Davis were developing another way to dig into the history of Europe via its DNA. They examined the genomes of a group of 2,257 Europeans, divided into forty different populations, and identified all the small segments of DNA that any two people had inherited from a recent common ancestor. The goal was to understand the way people are related in modern-day Europe and, in the process, to learn how human networks have changed through time.

Ralph and Coop found that any two modern Europeans who lived in neighboring populations shared between two and twelve genetic ancestors in the previous 1,500 years. These foreign cousins are a testament to the fundamental interconnectedness of all people but also to how ultimately disparate we are: Twelve genetic cousins in another country after 1,500 years isn't exactly grounds for a family reunion. A similar study found that in a sample of five thousand Europeans, there were tens of thousands of pairs of second to ninth cousins. According to Ralph and Coop, in the period between 1,500 and 2,500 years ago, a given pair of individuals would likely have shared one hundred or more genetic ancestors, which means they carried the same random segment of DNA passed down over all that time from an eightieth-grandparent—perhaps a Roman legionnaire, a Portuguese sailor, or a Greek shepherd. The farther apart people lived in Europe, the less likely it was that they shared genetic ancestors, although, say Ralph and Coop, they would still likely share some.

The way that genomes are cut, split, and shuffled across generations has significant consequences for their structure today. How do segments of DNA break apart and come back together again? I asked Ralph, who suggested that I re-create a lineage, at least on a tiny scale, to see how the process works. I returned to the kitchen table, and again sat there with my genome before me, only this time it was made of paper, a red strip to symbolize all the genetic material I received from my mother and a green strip to represent all the DNA I received from my father. I wanted to follow half of my DNA back through the generations, so I set the green paper aside, which left me with half my genome—everything I received from my mother.

I chopped the red strip into twenty-three smaller bits to represent the chromosomes in my cells that were made by my mother and then passed on to me. Then I passed the strips farther up the table, symbolically sending them back up the tree to my mother. In order to get a sense of what her genome looked like, I then added another twenty-three strips of brown paper to represent the DNA that my mother did not pass down to me.

The funny thing about the set of chromosomes before me was that even though I was looking at my mother's total genetic material, I was not looking at her actual chromosomes. The chromosomes she gave to me underwent a process of recombination before she passed them on. Segments of DNA were swapped between each pair, so that overall her chromosomes broke and then recombined on average about thirty-two times across the genome.

In order to unscramble the process, I chopped up the forty-six red and brown strips and swapped pieces between them. This was finicky work, so I settled for doing it just thirty times, which meant that most of the forty-six resulting strips had at least one red segment and one brown segment. These twenty-three pairs of mosaic strips represented my mother's actual chromosomes.

Because I wanted to follow my DNA back another generation to my mother's mother, I did the same thing for her. I discarded one chromosome from each of my mother's pairs, which left me with the twenty-three chromosomes she got from her mother. Again I pushed them a little farther up the table, back through time and to the spot in the tree where my maternal grandmother perched. Then I added twenty-three blue strips of paper to pair up with each of the chromosomes that my grandmother gave to my mother, to represent all the genetic material that my grandmother had but that was never passed down to my mother. In order to disassemble those twenty-three chromosome pairs and rebuild their original state in my grandmother's cells, I chopped them up and painstakingly pushed tiny little pieces of paper between the strips into alignment with others.

Some of the motley strips that represented my grandmother's original chromosomes had four different segments with three different colors. Because I was using specific colors to indicate where the DNA would end up, I could see which segments of my grandmother's DNA came down through the generations to me, which went to my mother but not to me, and which were not replicated at all. If I continued to do this for one hundred generations, all the pieces of my personal genome would become smaller and smaller and be dispersed further and further throughout my genealogical tree.

Yet as I dispersed and reconnected the chunks back through the time span of just a few generations, I could see they were not just different colors but also slightly different lengths. At first it was easy to divide the genomes up and push them back through the generations, but it wasn't too long before the chunks reached a size where they were not divided at all. They moved from one generation to another without changing in size. This was what Ralph wanted me to see: It is often the case that these segments of DNA will be passed on *whole* to the next generation and then, still whole, to the next. It may be many generations before they are once more chopped down.

A common misconception of the way DNA is passed on is that with each new generation any one section of DNA is divided in half before it is passed down. We have learned, of course, that there are exceptions. The genome isn't a perfectly smooth collection of equal bits that break up and come back together in exactly the same way. Because Y DNA and mtDNA don't get reshuffled with other DNA, they can be used to learn something about an individual in your family tree who lived 10,000, 50,000, or 100,000 years ago. That person is still there, in a sense, *in you* in a completely disproportionate way to the rest of your grandparents. The X chromosome is different too, as it recombines in an uneven way across the generations.

Still, despite the fact that we have come to accept the idiosyncrasies of Y and X chromosomes, as well as mtDNA, we have assumed that the rest of the genome followed the basic pattern of being halved, piece by piece, and halved again. As we have seen, the most impassioned antigenealogists like to cite the principle that if you go back ten generations, you have 1,024 ancestors in that generation alone, which means that the amount of genetic material you are getting from any one person in that group is 1/1,024 of their genome, so small as to be virtually meaningless. The implication is that you are a genetic soup, and it is pointless to look for patterns in soup. This is a powerful argument, but I have long wondered how much of its force comes not from its math but from its confident, almost cartoonish absolutism.

Indeed, according to Ralph and Coop, it turns out that many segments of autosomal DNA are moved *as is* from one generation to another and another without being reduced further. The reproduction of a genome is not a smooth process of disintegration into ever-smaller pieces; it's a lot more uneven than we suspected. As students Ralph and Coop were

taught the standard notion that because DNA is halved at each generation, it means you must get a quarter of your DNA from each grandparent, an eighth from each great-grandparent, and so on. When they first
realized this could not be true, Ralph told me, "It took us a while to get
used to the idea, because we were used to thinking in terms of a half, a
quarter, an eighth, because at first everything gets cut in half. But if you
take a segment of DNA and you cut it in half and you put it in two bags
and the next time you cut in half again, pretty soon the chunks that you
have are so short that the chance of them getting cut at all is really
small. So the whole chunk moves into one of the bags."

If every chunk of DNA were halved with every generation, the result
would be a rather neat picture of proportionately shrinking segments
that matched an expanding fan of cousins. But if the cut and shuffle of
DNA down through the generations is not a smooth, even process and
relatively large chunks of DNA may be passed on through generations
more or less unchanged, it has some interesting implications for what
DNA can tell us about the past. For example, if you share chunks of
DNA with a fifth or, let's say, a twentieth cousin that have been inherited
from a common ancestor, those chunks may be the same size as each
other. Ralph suggests the window of relatedness is about 500 years long,
meaning that if you share a chunk of a certain length with someone, you
probably had a common ancestor somewhere within the last 500 years.
A smaller chunk could have been inherited from an ancestor who lived
anywhere from 500 to 1,000 years ago. An even smaller chunk could be
from a common ancestor who lived between 1,000 and 1,500 years ago.

This has implications for the genetic databases at companies like
Family Tree DNA and 23andMe. Users may find relatives who are first,
second, or third cousins based on how much DNA they have in common.
They will also find they belong to a massive group of more distant
cousins. According to Ralph, based on the DNA you have in common,
these distant cousins may be fifth cousins, but they might also be fifteenth cousins, and you may share a common ancestor much further
back than you think. In fact, when you consider any pedigree, said
Ralph, you soon stop getting many new chunks. We may be made anew
in each generation, but it's out of blocks, not powder.

There are other ways that the transmission of DNA over many generations is irregular and that may have a big effect. For instance, women have a higher rate of recombination than men. The size of the population that your ancestors came from and how stable it was affect what is passed down. Keep in mind too that thirty-two is just the average number of times the chromosomes break and recombine at each generation, and that can vary. Geneticists also talk about "hot spots" on chromosomes, places that may be more inclined to recombine over and over.

When you take all of this into account, you can use segments of identical DNA from common ancestors to unearth connections from the past three thousand to four thousand years.

Curiously, even if some long-past relatives are more prominently represented in our genome than we imagined (relatively speaking), there are many more who have simply evaporated. This is because our personal genetic tree is not equivalent to our genealogical tree, which is to say that not every one of our direct ancestors has contributed to our genome.

If you look back ten generations, everyone in your lineage is by definition a member of your genealogical tree. The fact that they paired up with the spouse they did and then had a child who became your ancestor is an unchangeable, existential bond. When you first start looking back into your lineage, the genetic and genealogical trees are, of course, exactly the same: Your *genealogical* parents are your *biological* parents, and so on. But there is a point in time at which the genetic connection between you and most of your many generations of grandparents vanishes away to nothing, despite the fact that your genealogical tree keeps growing.

Beginning around eight to ten generations back, geneticists agree, there were so many people contributing their DNA to your family line that by the time you came along, a lot of it has simply dropped out. If you could trace every piece of your DNA back through time, it would follow the branches of your genealogy, but it would trace fewer and fewer of those lines as it went on, leaving your genetic tree much smaller than your genealogical tree and leaving you in the odd position of being biologically unrelated to many of your blood relatives.

Estimates of how many people drop out of the genetic tree vary. By the time you go back ten generations, it may be that you have completely lost the genetics of at least one and likely many more grandparents. By the time you go back sixteen generations, you will have 65,336 ancestors (not counting possible repeats where an ancestor may appear on more than one lineage). According to Nick Patterson at the Broad Institute, this is far more than the number of distinct chunks of ancestry in your genome from that generation; there will only be around one thousand such chunks contributed by a sixteenth-generation ancestor. So most of your sixteenth-generation ancestors will have contributed nothing to you genetically. If you built a time machine and traveled back four hundred years and, let's say for the sake of argument, found yourself in a romance with one of your sixteenth-grade grandmothers, the good news is that you can feel fine about having children together. However morally bizarre that might be, it would not be genetically problematic.

But wait, doesn't this mean the antigenealogists have a point—isn't this the same thing as saying that the human genome is quickly chopped down into meaningless dust over the generations? While it's true that a lot of our personal genome eventually dissolves, it does so in a patterned way and there is meaning in the trail it leaves. If some people disappear from our genome while others remain, that is one more pattern shaped by history.

If you take any two modern Europeans, no matter how far apart they live, they will likely share millions of genealogical, if not genetic, ancestors within the last thousand years. Still, even though the genealogical tree is much bigger than the genetic tree, it doesn't fan out forever either. Because the number of ancestors in a tree doubles with each generation, any person's tree of possible ancestors grows exponentially and quickly reaches a point where it is greater than the population of the world at the time.

For example, geneticists and historians vary somewhat in their estimate of how long a generation is. Some say twenty years and some say thirty, so let's assume it's around twenty-five years. We'll also assume that there is just one person in the world today, let's say *you*. In order to arrive at you, there must have been more than a billion people on the planet approximately 750 years ago. Or, to put it another way, about thirty generations ago there would have to have been roughly 2 billion people in the world, whose children and grandchildren and so forth met one another and married until one day you appeared. But in fact there were only 400 million people in the world 750 years ago. What this means is that your thirtieth-great-grandparent along one line is probably your thirtieth-great-grandparent along many other lines too. Genealogists call this pedigree collapse.

Pedigree collapse is common in family trees that go back to the nineteenth century and earlier. In many different cultures marriage between cousins was not the exception but the rule. (For more on the demographics and implications of cousin marriage, see chapter 14.) Some scientists, like Ralph and Coop, argue that by two thousand to three thousand years ago, everyone who was alive at that time across the globe was actually a genealogical ancestor of everyone alive today. The argument is a purely mathematical one based on the idea that, because your theoretical genealogical tree would be so massive two thousand to three thousand years ago (over 120,000 trillion ancestors), it must surely include all of the much smaller number of individuals (50 million to 170 million) who were alive in the world at that time. Many population geneticists I spoke to found this to be a completely noncontroversial theory. It wouldn't take long for the networks of relationships in different countries to be altered by the introduction of just one or two travelers from a distant area, connecting all of the populations of the world. Others whose work is more engaged with the events of history found the idea implausible: While it is technically possible that the connecting events happened, they believe it is more likely that the world's populations were more completely isolated from one another for a longer period.

If the hyperconnectedness of humanity is true, it would mean that everyone alive today—you, your neighbor, Vladimir Putin, and the emperor of Japan—could count the same Egyptian pharaoh, as well as everyone else alive at the time, as a distant grandparent. A set of common genealogical ancestors doesn't mean, of course, that people today don't have genetic differences (which can be seen in any number of experiments, such as those exploring the genetics of Britain described in the previous chapter). It doesn't mean either that once you go back three thousand years everyone's family history is effectively the same. Even if both you and the emperor of Japan can count the same pharaoh in your family tree, the pharaoh may appear many more times in your tree than the emperor's. Genetic history would not be possible if everyone's genealogical tree were identical a few thousand years ago. If you imagine a great network of ancestors stretching from now back through time, with every person a node in the network, all the people alive in the world three thousand years ago who left any descendants would be significant nodes, because each person alive today could trace a path back to them—but it wouldn't be the *same* path. Some people would trace thousands of paths back to the same node, while others would trace many fewer.

Curiously, it is often the case that when someone says, "We are *all* related to Confucius or Boadicea or Erik the Red," the implication is that there is no texture in this history, that if we look back far enough, everyone in our family was the same, so there is little of interest to say about the connection that one person or group of people may have with people from the past. This is not the case. The topology of the human network, in which we are all nodes, is incredibly complicated. While there are points of sameness—perhaps we can all trace at least one connection back to everyone from three thousand years ago—that sameness does not mean that all the other pathways we trace back to shared ancestors do not have great significance. If you start tracking the ancestry of segments of the genome, like the Y chromosome and mtDNA, the picture of shared ancestry becomes much more complicated again. The patterns of all the paths tell us things about history that we might otherwise never know.

Recall that in the British genetics project people in Cornwall and Devon and other regions carried a signature of their local region from a certain time. The fact that they had a Cornish or Devonian flavor to their genome doesn't mean that they didn't also have a few ancestors from Scotland or France or that there weren't Catholics or Vikings or even a Chinese philosopher in the genealogical mix. Genetically they were not *purely* one thing; the same is true of their genealogy. Their heritage was a matter of dilution—enough of their ancestry had come from the one time and place that it was still recognizable. This will be true whether we look back one thousand years or five thousand years.

In addition to some general rules of relatedness in Europe, Ralph and Coop identified some tantalizing variations. People in southeastern Europe seem to share many more genetic common ancestors than people in other parts of Europe. The period when they share this large cluster of common ancestors dates to around 1,500 years ago, when there were significant Slavic and Hunnic expansions.

In dramatic contrast to the rest of Europe, most of the common ancestors of Italians seem to have lived around 2,500 years ago, dating to the time of the Roman Republic—the period that preceded the Roman Empire. Modern Italians certainly share ancestors within the last 2,500 years, but far fewer of them. In fact, Italians from different regions of Italy today share about the same number of common ancestors with one another as they do with people of other countries. It is as if other European nations are more genetically homogeneous, while Italy is composed of many smaller countries. How do we make sense of this?

To understand it we have to understand not only the way that the living material of families is passed on but also the world in which the families were formed. While DNA was spread across Italy by all the normal processes of reproduction, those processes were shaped by the culture and geography in which they took place. Genomics by itself won't give us the whole picture of European history.

There are many stories that might explain the peculiar pattern of Italian common ancestry. For example, Ralph and Coop suggest that the region may have been less affected by the population expansions that spread across the rest of Europe in the last two thousand years. To test the possible scenarios, said Ralph, they would have to look at demographic, linguistic, and other historical patterns. Many factors would have to be taken into account, such as whether the difference in common ancestors was true for people whose families had lived in Italy for centuries (which would confirm that the effect did not exist simply because an entirely new population moved in from a different country five hundred years ago).

I was intrigued by the way that, in addition to the larger group of common ancestors from 2,500 years ago, other genetic analyses of Italy show that many small different genetic groups are related to each other in a way that forms a gradient from north to south in the country. I asked Guido Tabellini, the economist who measured the difference in social capital between the north and south of Italy, if he could think of a reason why there were fewer common ancestors between some groups in the most recent 2,500 years. He pointed out that the political entities that ruled northern and southern Italy were separate until the middle of the nineteenth century, and that separation reduced economic and social integration within the country as a whole. In addition, he said, the north and south were ruled in very different ways: "Northern Italy had the tradition of more independent and enlightened government, whereas southern Italy had the tradition of being dominated by external influence.

"A second important difference which has not been studied much," said Tabellini, "is the role of the family. If you look at the family traditions within Italy, they are also very different, and so are attitudes towards females. The attitude towards people outside of the family can also be explained by these different family traditions."

It is possible that the social currents that Tabellini describes could help explain the unique pattern of ancestor sharing in Italy. It is possible that the pattern may be explained by an entirely different story. To find the best story, geneticists, historians, and other researchers will have to work together.

As dazzling as the possibilities of genetic history are, it wasn't the possibility of uncovering realities that were once completely invisible that motivated some of the first genetic genealogists. James LeVoy Sorenson died in 2008, and when I asked Woodward why he and Sorenson did what they did, he leaned back in his chair and, making air quotes, said one of the big goals was a "'Miss America' kind of goal. World peace. That kind of thing."

Woodward said that the men had dreamed that, once their project was completed, they would be able to take any two people in the world, sit them down, show them a piece of DNA that they shared, and say, "Here's the common ancestor that you share."

As he and Sorenson went back and forth, Woodward recalled, they thought "perhaps if people understood and knew how closely they were related to each other, they would treat each other differently, hopefully better."

Chapter 11

The Politics of DNA

By nature, men are nearly alike; by practice, they get to be wide
apart.

—Confucius

For almost two hundred years the official biography of one of America's founding fathers, Thomas Jefferson, stated that the philosopher, statesman, architect, president, writer of the Declaration of Independence, and designer of the graceful manor at Monticello was married to Martha Wayles. Jefferson and Wayles had six children, but only two daughters survived to adulthood. Jefferson documented his own family history back to Great Britain; his father's family was from Wales, and his mother's family had come from England and Scotland. Recall that when he discussed his pedigree in his autobiography, he added that the reader should ascribe to that information "the faith & merit he chooses."

Yet even in Jefferson's lifetime another story competed with the accepted version. In this tale Jefferson's family included many more children than those of Martha Wayles. It was alleged that after the death of his wife, Jefferson began a thirty-eight-year-long liaison with his slave Sally Hemings, who herself had been born to a white father and a mother of mixed race. The story was reported in the press in Jefferson's day, beginning with the claim that Jefferson and Hemings had a son, Thomas Woodson, who was sent to live on another estate when he was twelve. It was also said that Jefferson fathered the six children that Hemings raised at Monticello.

The most compelling evidence for the Jefferson/Hemings relationship exists in the testimony of many people who lived at Monticello at the time and passed the story down to their children. Strong and detailed oral histories were kept by generation after generation, especially within the families that descended from Hemings, all testifying to the fact that Jefferson had fathered Hemings's children. Even in the late twentieth century, distant Hemings cousins who had never met told similar stories about the day their parent pulled them aside and whispered to them that one of America's most venerated founding fathers was also more directly one of their own.

Up until then some scholars had been willing to concede that Hemings's children may have been fathered by Jefferson's nephew, but not by the great man himself. For most the official story held sway. Its power, legal scholar and historian Annette Gordon-Reed argued, derived mostly from the fact that historians tended to use a simple rule to evaluate evidence: The word of white people was good, and the word of slaves was not.

In 1997 Annette Gordon-Reed published *Thomas Jefferson and Sally Hemings: An American Controversy*, one of the first books to systematically examine the evidence for and against the Jefferson/Hemings claims. She concluded that the relationship did exist. Many critics reacted as if she had launched a personal attack on one of the nation's most beloved historical figures. Others were saddened but dismissive. Gordon S. Wood, Pulitzer Prize–winning author and professor of history, wrote:

> This idea of an intimate and loving relationship between Jefferson and his black slave may have gained great power and increasing credibility in our culture because it represents the deep yearnings of many Americans; it symbolizes what many of us believe is the ultimate solution to our race problem.
>
> If only it were true. . . . But wishing won't make it a historical reality.

Two years later Eugene Foster, a retired professor of pathology who lived in Charlottesville, Virginia, realized that the new genetic genealogy meant the Y chromosome of Hemings and Jefferson descendants could throw light on the matter. He collected four samples: one from a direct male descendant of Jefferson's paternal uncle (while there were no living direct male descendants from the Jefferson/Wayles marriage, this Y would be the same as Jefferson's); one from a direct male descendent of one of Jefferson's nephews by his sisters; one from a direct male descendant of Eston Hemings, Sally's son; and one from a direct male descendant of Thomas Woodson.

Foster found that the Y chromosome that had been passed down from Eston Hemings matched the Y chromosome passed down from Jefferson's uncle. Because the Y was an especially rare one, the match was powerful evidence that Sally's son Eston was also a Jefferson. The old rumor that Jefferson's nephew fathered Sally Hemings's children was also quickly dispatched, as the Y from the nephew's descendants did not match the Hemings Y.

Because the Jefferson Y was shared by Jefferson's other male relatives who are known to have visited Monticello at the time, the case that Thomas Jefferson himself was the father of Hemings's children was still a circumstantial one. Yet the circumstances were overwhelming—not just the DNA evidence but also the persistent rumors, the detailed oral histories, the preferential treatment that Jefferson gave to Hemings's children (not just Eston but also his brother Madison and their other siblings), and careful analyses of the times when Jefferson and other men were present at the estate.

Well-known geneticist Eric Lander and historian Joseph J. Ellis wrote at the time that the "burden of proof has clearly shifted" to those who would deny the Jefferson/Hemings link. The Thomas Jefferson Memorial Foundation appointed a research committee, which found that the "best evidence available suggests the strong likelihood that Thomas Jefferson and Sally Hemings had a relationship over time that led to the birth of one, and perhaps all, of the known children of Sally Hemings." Historians who had long argued that the Hemings story was a myth accepted that they had been wrong, and for many people who witnessed these dramas only from afar, the twist in the epic tale of race, power, and class in America was cause for celebration. There was a sense that thanks to DNA, the actual lives of historical figures—or anyone else, for that matter—could no longer be hidden behind a wall of respectability and lies.

Foster also tested the Y chromosome of the Woodson family. The Woodsons had a powerful oral history linking them to Hemings and Jefferson, and of all the families connected to the story, they had been the most public with their claim and were the most confident that the Y-chromosome test would prove what they had long believed to be true. As early as 1978, at the first Woodson reunion, the family had discovered that many members from lineages who had never been in contact had all kept the same story alive. Since the turn of the nineteenth century, Woodson descendants had been passing down the story that Jefferson was Woodson's father. Even Eugene Foster, who set the DNA test up, said that he was expecting the DNA to validate the Woodsons' claim.

On its Web site today the Thomas Woodson Family Association states that Thomas Corbin Woodson was "the issue of a union of Thomas Jefferson and his slave Sally Hemings." But that is not what Foster's test showed. The Y-chromosome test that linked Thomas Jefferson to the paternity of Eston Hemings also showed that the Woodson family were not his descendants.

Michele Cooley-Quille remembered when she was twelve years old being told by her father, Robert Cooley III, the first African American federal magistrate, that she and her two siblings were the great-great-great-great-great-great-grandchildren of Thomas Jefferson. "We were so excited," Cooley-Quille told a reporter many years later. "It's so exciting to realize that part of the blood that runs through your veins was Thomas Jefferson's."

Cooley-Quille's father explained that her fifth-great-grandfather, Thomas Woodson, was the first child of Jefferson and Sally Hemings. In 1998, when Cooley-Quille was interviewed about the family story, she was a professor of psychology at Johns Hopkins School of Health, a member of the Thomas Woodson Family Association, and pregnant with her first child. She said that she would pass her family story along to the next generation. "Having a strong sense of family is so important," she said, "and I think we undervalue it. It gives a sense of how to operate in this world and how to operate positively."

When Foster's Y-chromosome results were published in *Nature*, the

Woodsons were devastated. According to Sloan Williams, a biological anthropologist who became involved with the family, their first response was disbelief, and they wrestled painfully with the gap between the long-held stories and the DNA evidence. Overall, Williams explained, there was incomprehension that the oral histories could be wrong. "They matched so well. They came from independent sources. The family couldn't understand why Thomas Woodson would have claimed Thomas Jefferson as his father. He didn't gain anything. He risked things." In an account of her experience at the time, Williams wrote, "They were extremely suspicious of the results and of the researchers who performed the work."

Not only was the family taken aback by the test result, but they were also angry about the way the results had been announced. Foster had assured them he would let them know what he found before it was published, but the news was leaked to the press before the family was informed. Robert Golden, the head of the Thomas Woodson Family Association, learned of it when *U.S. News & World Report* phoned him and asked him how he felt about the DNA result.

In response to the announcement and the ensuing media onslaught, the Thomas Woodson Family Association formed a research committee to investigate the experiment. Because one Woodson family member, the historian Carolyn Moore, was a colleague of Sloan Williams, she asked her for assistance. Moore's first request was that Williams simply help the family understand the genetics. Williams expected this would take a lunch, followed perhaps by a phone call or two. But when they met, Moore pulled out an enormous tome, "The Woodson Source Book," which contained copies of all the documents pertaining to the family's history. It became apparent to Williams that her involvement was going to take a while.

Explaining the basics of the genetics tests and whether they were valid required many sessions. In 2000 Williams traveled to a Woodson family reunion and met with members of the research committee and other family to discuss the case further. The committee representatives asked her whether mutations could explain the results (they couldn't) and if differences between the Y chromosomes had been appropriately interpreted (they had). Their main goal, according to Williams, was to find plausible alternative explanations for the results.

But none of the alternative explanations were appealing. As Williams pointed out, the descendants of Thomas Woodson's two oldest sons shared the same Y, which meant that the two sons themselves shared the same Y, which in turn implied that Thomas Woodson himself, their father, had that Y. The fact that this Y did not match the Jefferson Y suggested that Thomas Woodson was not fathered by Thomas Jefferson. If, for the sake of argument, you accepted that Thomas Woodson *was* fathered by Jefferson, then the only explanation for the Y shared by Thomas Woodson's two eldest sons would be that they were not actually Thomas Woodson's sons. Perhaps they were Woodson's wife's sons from a previous marriage? If that were the case, these Woodson descendants would not be able to claim either Thomas Jefferson or Thomas Woodson in their ancestry.

The Woodson family asked Foster to also test the Y chromosome of a descendant of Thomas Woodson's younger son. The test confirmed that the Woodson sons shared the same chromosome, leaving the modern Woodsons with the same dilemma: Either that Y was Thomas Woodson's or it was someone else's. In either case it was not Jefferson's.

Since its first generation the Woodson family has included many admirable, talented, and strong leaders, including Lewis Woodson, a preacher and abolitionist, once called "the father of black nationalism." "The Woodsons were understandably proud of their successful and accomplished family," wrote Williams. "The loss of oral history that they had all shared and passed down from generation to generation shook their belief in the qualities and values that their family represented."

By the time of the next family reunion, some Woodsons had begun to grudgingly accept the DNA results, but others refused to believe. Williams wrote that many family members took a similar position to that of the head of the association, who recalled that Woodson descendants from all over the United States had shared the same oral history when they met for the first time in 1978. "So I don't care, you know," he stated. "I have respect for the DNA study and et cetera, but I still wonder. . . . It hasn't changed my belief at all."

According to Williams, Michele Cooley-Quille was vehement in her rejection of the results. Yet Williams carried out an independent test and she confirmed Foster's finding. Byron Woodson, Cooley-Quille's brother, suggested at the time that the DNA results had been tampered with. Such is the jumbling of genetic material when anyone reproduces that siblings may look quite alike or quite different. Even though they were full brothers born within three years of each other, Madison and Eston Hemings looked different enough that when Eston moved his family to Wisconsin, he changed his surname to Jefferson and from then on identified as white. Madison's family stayed in rural Ohio, and while it was said that some family members had "disappeared into white society," many remained in African American communities and identified as black. Byron Woodson pointed out that the only confirmed connection to Jefferson was with the descendants of Eston, who had moved into white society long ago. He attributed this to racism. Still, the claim confused Williams, given that the men in the Eston Hemings line were confirmed descendants of Jefferson, and by implication the men in the Madison Hemings line—many of whom identified as black—were as well.

The Woodsons then located some descendants of the owner of the estate on which Thomas Woodson grew up. Although the Woodson Y did not match the Jefferson Y, it was notable that the Y was a type that was associated with a European heritage. As was common at the time, young Thomas took his surname from the estate master, John Woodson. The family asked Williams to test the new Woodson Y. She found that the descendants of John Woodson shared the same Y, but it was different from the Y of the Thomas Woodson family. The Woodsons then discussed the possibility of locating the graves of Thomas Woodson and Sally Hemings with a view to running a DNA test on their remains, but the exact location of neither grave could be found. Later a family member had his DNA tested with a genetic genealogy company and asked Williams to help him understand the results. They confirmed what was already known.

The Jefferson/Woodson story was painful because a family that had achieved so much in the face of adversity felt that they had been ejected from a larger group to which they proudly belonged and from a history that belonged to them. The Jefferson story had given family members great strength and motivation at significant moments in their lives.

The consequences for the lineages of Madison and Eston Hemings, both black and white, were different: For them the DNA provided a triumphant vindication of their family history. It was not an abstract vindication, either. As Williams pointed out the story of Madison and Eston's paternity changed the lives of individuals within those families, and it provided the rest of the country with a newly accurate model of how some postrevolutionary families were shaped. The lineages of Madison and Eston Hemings had lost touch with each other, but when the Jefferson DNA results were published, Julia Jefferson Westerinen, a white woman and a descendant of Eston, and Shay Banks-Young, a black woman and a descendant of Madison, met for the first time. Since then, Thomas Jefferson's white and black granddaughters have publicly spoken many times about the way that embracing each other and learning from each other has changed their lives.

It's likely that as we become more adept at reading the stories of the past in the molecules of the present, this new knowledge will affect *whom* some people feel they belong to. When Foster analyzed the Jefferson Y, genetic genealogy was an infant science, and companies like the Sorenson Molecular Genealogy Foundation and Family Tree DNA were in their earliest stages of development. It took a private researcher like Foster, with his background in pathology, to devise the experiment, order the analyses, and interpret the results. Less than fifteen years later, anyone with a credit card can dip into his family's invisible history via his own genome.

The power that these tests give us—to see with such clarity so far back into the history of humanity—is unprecedented. The additional opportunity to locate an individual's personal history within that larger context was, up to the point where it became possible, completely unimaginable. Yet this new knowledge does not come without cost. Users risk discovering facts that for whatever reason they may not wish to know. The only way to truly block such risks is to halt the research or to legislatively restrict access to it. If someone could have stopped the

Jefferson study before it changed history, it would have meant that on the one hand, the Woodsons would have been able to keep their legacy, but on the other, all the descendants of Madison and Eston Hemings would have continued to be disavowed.

Despite its potential for great good, many critics have spoken forcefully about the negative impacts of reading history in DNA. The concerns are real, and much of the commentary reflects a sense of responsibility in the scientific community. Still, the collective response can give the impression of a great, looming threat.

In 2007 a group of scientists published a policy discussion about ancestry testing in *Science*, in which they stated, "Genetic ancestry testing . . . has serious consequences. Test-takers may reshape their personal identities and they may suffer emotional distress if the results are unexpected or undesired." The American Society of Human Genetics examined the topic with some optimism but much hand-wringing in 2010, noting that "the very concept of ancestry is subject to misunderstanding in both the general and the scientific communities." An earlier paper in the *British Medical Journal* stated, "Tracing genetic identity can lead to the resolution of uncertainty but can cause more problems than it solves." Some critics view the motives of anyone who wishes to look at his own genome as potentially suspect.

Writing about early-stage misinterpretations of genetic data, the biological anthropologist Jonathan Marks worried, "With genetic data, it seems, one could find entities that did not really exist, or impose cultural assumptions on the data and mistake them for patterns inherent in the data, yet still cloak oneself unimpeachably in the mantle of modern science." In a caption to a racist depiction from 1842 of the different skull shapes of a Caucasian, an African, and a chimpanzee, he wrote, "As the authoritative voice on identity and descent, science's track record is hardly blemish free"—a sentiment with which few scientists would disagree.

Kim TallBear, an assistant professor of American Indian studies at Arizona State University, wrote that genomic scientific techniques, specifically those of the Genographic Project, have emerged out of the racial science dating back to the seventeenth century. The hope expressed by Genographic and genetic genealogy companies that this science was different—that it sought not to affirm racist categories but to demonstrate that they were cultural and unsupported by nature—was, to her mind, "naive at best." TallBear also wrote that the claim that genetic proof of humanity's African origins was "anti-racist" was complicated by the fact that it involved "portraying Africa and Africans as primordial."

The intent, of course, is to ward off the most venal use of biology, the specter hanging over all genetic research and all genealogy as well. Men like Francis Galton, Madison Grant, Heinrich Himmler, and Adolf Hitler have left in their wake a deep fear about the abuse of science, including the notion that merely considering the human genetic network will either make us all racist or justify our latent racism. I once spoke about the genetics of ancestry with a Holocaust historian who had hunted some of the last surviving Nazis in the 1990s. When I told him that little letters in our genetic code might testify to the ethnicity of our parents and grandparents, he said, "The Nazis would have loved this." They would certainly have seized upon the idea, but in the end the full picture would have let them down just as badly as all the other dubious measures of race they tried to develop. Comparing the volumes of people's skulls proved pointless, as did trying to formulate an objective measure of beauty. Likewise, the bits of DNA that genetic historians study do not indicate what a person will look like, think like, or live like. They are records only of ancestry; they tell us that groups once existed that, for whatever reason, lived together long enough that they ended up with genetic commonalities.

Around the same time that the Jefferson research was taking place, another hugely ambitious genetic project was stalling spectacularly. In the early 1990s, Luigi Luca Cavalli-Sforza founded the Human Genome Diversity Project (HGDP), the goal of which was to rebuild the biological and language trees of humanity by sampling the DNA of tens of thousands of people from all over the globe. It would be hard to overstate the contribution that Cavalli-Sforza has made, not just to the science of population genetics but to the imaginations of generations of population geneticists and historians. His early projects in the 1960s were the first attempts to rebuild the history of the world through the distribution of traits found in blood. His popular book, *Genes, Peoples, and Languages*, first published in 2001, outlined his ambitious work, bringing together evidence from languages and genes to unearth the history of humans. The Human Genome Diversity Project would not only enrich history, argued Cavalli-Sforza, but would also have medical applications and, by demonstrating that there was no such thing as biological race, would be a tool against racism.

For all its apparent idealism, the project's organizers did not pay enough attention to addressing the contexts in which the project's requests for blood, and ultimately for knowledge, were made. It was one thing for middle-class westerners who have access to the educational and medical outcomes of the project to share their blood, and the information in it. But many indigenous groups of interest to the project were still fighting for basic human rights and health care. They struggled with extreme poverty, terrible health, and overrepresentation in the criminal justice system. Many had experienced exploitation and even medical experimentation at the hands of scientists.

In response, small, well-organized activist groups created many impediments to the progress of the project. They pointed out that it was unclear whether pharmaceutical companies would be able to access the DNA and not only use the information it contained to make enormous profits but even to patent people's genes. They claimed that some indigenous groups were afraid that the scientific story of the world would be used to rewrite their own cosmologies. Their objections were connected to issues of genetic ownership too, such as the question of whether an individual had a right to share his own DNA with the project if other people in his community did not want to. Ultimately, the contradiction that millions of dollars were being invested in the HGDP but not in the people whose blood it would use became overwhelming. Among some indigenous groups it became known as the Vampire Project. (At the same time, many of the same political battles were being fought over genetic and other research conducted on ancient remains.)

A small group of cultural anthropologists accused Cavalli-Sforza and his colleagues of racism and hubris. Some even questioned the scientific merit of the project, suggesting that its questions about human history were obscure and asking whether the genetics of a single group would ever shed light on any history but its own. Years later in a United Nations address, Cavalli-Sforza said, "Ignorance can breed fear and hate, but I have discovered that it is most dangerous when mixed with the personal political agenda of science haters."

For obvious reasons, genetic research with subjects whose history is less troubled provokes less overt concern. Recall that the genomes of the people in the British genetics project were so overwhelmingly similar that as far as current medical genomics is concerned, they were effectively the same. Yet there were still identifiable differences that told a historical tale: The subjects' ancestors had lived in different regions and they had left a mark in the genetics of their modern-day descendants. As far as we can currently tell, these differences were shaped by the same neutral evolutionary mechanisms that result in some groups having differently shaped skulls.

What is perhaps most confusing about the criticism of this kind of genetic research is that its detractors often cite one of the most popular ideas of the human genome era: namely, that DNA reveals that race is a myth and that beneath the skin we are all fundamentally the same. But how can this be true when another consequence of the human genome era is that we can now have our genome analyzed and our racial history quantified? Does race exist in our genes or just in our heads?

The modern notion that there is no such thing as biological race and that we are more alike across populations than within them can be traced to 1972, when Richard Lewontin, an evolutionary biologist and geneticist at Harvard, conducted a landmark experiment that continues to shape how people think about the subject. Lewontin looked at seventeen places in the genome where a single letter might be different between people, and he showed that for each of these spots there were more differences *within* populations—or what we often think of as racial groups—than *between* them. The implication was that the largest differences that exist between people occur along all sorts of spectra, none of which is racial. As Lewontin wrote:

Human races and populations are remarkably similar to each other, with the largest part of human variation being accounted for by differences between individuals. Human racial classification is of no social value and is positively destructive of social and human relations. Since such racial classification is now seen to be of virtually no genetic or taxonomic significance either, no justification can be offered for its continuance.

These findings were generally taken to mean that the ways we differ from one another across racial or ethnic divides are far, far smaller than everything we have in common; that even though individuals from different parts of the world may look different from us, they are generally going to be more like us than not; and that there is no biology of race. The findings were also taken to mean that you cannot identify a person's background from his DNA. The Human Genome Project announced that "two random individuals from any one group are almost as different as any two random individuals from the entire world."

However, if you look at even more bits of the genome, the picture changes. For example, in 2007 a team led by D. J. Witherspoon, a researcher at the University of Utah, examined the same question. They confirmed that if you were comparing a few hundred bits between people across traditional racial divides, you would still find that many individuals may have more in common across the divide than within their own group. However, if you took people from different populations in different parts of the globe and you compared *thousands* of bits of DNA, the picture changed: Increasing the resolution by looking at more DNA shows that people will tend to have more in common with an individual from their own population than from a distant one.

Witherspoon and his colleagues refined the approach even further and demonstrated that actually, when comparing populations that have been separated geographically for a long period of time, you need only a hundred bits of DNA to tell which person came from which population, which may make it seem as if race could be detected in DNA. Actually, the researchers did find something in this data, but it was *not* race.

There are many reasons for this, some of which have nothing to do with genes. The first reason they didn't find race in the genome is because it was never there to be found; *race* is an imprecise and ultimately unhelpful notion in biology. Part of its power comes from the implication that the divisions it refers to are absolute and eternal, yet "race" is one of the shiftiest words in language. Sally Hemings's mother had a white father and a black mother, and Hemings's father was white (Hemings shared a father with Martha Wayles Jefferson, the president's wife). Yet had Hemings been included in consecutive censuses since 1790, when then–Secretary of State Thomas Jefferson instituted the first American census, at first she would have been noted without name or race, and she later may have been classified at different times as "mulatto," or "black," or "white." The collection of traits that are supposed to distinguish different races changes in different eras, depending on who has power and who doesn't. Not only is race defined with a good deal of arbitrariness, but *who* gets to define it changes too. Sometimes a racial category is imposed on people, and sometimes it's one that people choose for themselves. Race fuses cultural traits with physical traits, and it presumes either that what is cultural is determined by what is physical or that they always go hand in hand.

Again and again science has shown through both its failures and its successes—from the grotesqueries of eugenic "science" and failed Nazi attempts to quantify race to the positive revelations of the Y chromosome— that the categorical boundaries we draw between people when we talk about race are always in part culturally determined; they never exactly fit onto real populations. There is simply no predetermined set of genetic or other physical divisions into which different human groups throughout space and time can be discretely assigned. Modern-day racists may wish to believe that some DNA is more privileged than others, but nothing in the human genome can be explained by the age-old foils of racism, such as platonic intelligence or beauty or purity.

Still, as unhelpful as the idea of race is, it is a hard one to shake off. Simply asserting that race does not exist doesn't appear to be changing people's minds or lives, perhaps because it seems to be so flatly contradicted by their more vivid daily encounters with different groups of people that look quite dissimilar. Insisting that race is a cultural construct doesn't help people understand the common experience of meeting a person who appears to be, say, Chinese or northern European and finding out that she was indeed born in China or northern Europe. If race is not the thing we see in other people, then what *is* it?

The confusion arises because when we use the term "race," we often include the idea of *ancestry*. This becomes a problem when people who want to reject race, or at least prove that it has no biological underpinning, effectively reject the idea of ancestry too. Early on critics attributed interest in the genetics of ethnicity to an unmoored "faith" in genetics as a solution to disease. In response to a 2005 *New York Times* op-ed about the medical utility of "race," one political scientist argued that staff and grantees of the Department of Health and Human Services and National Institutes of Health should not publish or cite anything suggesting that genetics is associated with any population category, including nationality or ethnicity, unless the finding was statistically significant *and* the description would "yield clear benefits for public health."

While these proposed guidelines are extreme, they exemplify a widespread anxiety that is not often so boldly articulated. Declaring statistically significant information off-limits by fiat is unscientific, and it has worrying implications for free speech. Far more important, the measure would endanger public health more than protect it. Studies of genetic correlates of disease are easily confounded by markers of ancestry. Medical research that hopes to identify genetic causes will risk being misled by false positives if it ignores the ideologically neutral markers of ancestry.

Ancestry is real and it can't just be defined away. You can see it on people's faces, and you can definitely identify it in their DNA. The *it* that makes letters of the genome fall into different patterns in different groups is in fact the *ancestry* of the people carrying them.

In 2011 Eran Elhaik was hired to solve one of the biggest jigsaw puzzles in the history of the human race; in fact, the puzzle was the history of the human race itself. Following the failed Human Genome Diversity Project, National Geographic launched the Genographic Project in 2005 to develop a way of reading people's Y chromosome and mtDNA, and half a million people eagerly participated by contributing samples of their DNA. From the beginning, indigenous communities were approached in a different way. It was made clear that they retained ownership of their own DNA. Whereas the HGDP had proposed to keep the cell lines from samples alive in perpetuity, the new project committed to not do that, as the idea that cells would live on after their owners' deaths was disturbing for many groups. A mouthwash was developed for sampling so that people who objected to the idea of giving blood did not have to do that. While it is hardly the case that the problems of indigenous groups have been solved, there is generally a more mutually respectful and appreciative relationship between this project and the groups it engages with.

In 2012 Genographic decided to include all the chromosomes and analyze autosomal DNA as well. Elhaik was asked to design a method that would extract the most information from a sample but at the same time extract only historical information and not anything to do with an individual's health or features. (See chapter 14 for more about health and the genome.) Before he could do that, Elhaik had to collect a big enough set of data to survey as many different populations as possible, as he needed to see the whole in order to understand its parts. The trickiest part of the challenge was that whatever series of letters made any two populations distinct from each other was probably going to be a different set from the series that made any other two populations different.

Elhaik spent years obsessively collecting data. He collected as much as he could from publicly available data sets, and he was given data as well. "There were a lot of data-rich scientists who were kind enough to share their data with me," he recalled. Ultimately he was able to gather the genetic data of tens of thousands of people from almost five hundred different populations, amassing the largest set of its kind in the world.

Elhaik worked out how all the populations differed from one another by comparing each group to another one by one and working out the minimum number of letters he needed to be able to distinguish the pair. "If you have a Lebanese and a Syrian, you ask, Do I need a hundred, two hundred, a thousand, two thousand genetic markers so I can correctly classify a Lebanese as a Lebanese and a Syrian as a Syrian?"

"You cannot do it for every population," Elhaik clarified, "because some of them are genetically indistinguishable. I had a lot of Indian groups, including different linguistic groups and castes, but no matter how many genetic markers you are going to use, they were not separable." Elhaik found that he needed between five hundred and two thousand letters of DNA to tell most of the subject groups apart.

Elhaik's subject populations roughly corresponded to what we think of as different ethnic or racial groups, but it was actually ancestry that he measured. This is not a semantic trick, an attempt to replace an incendiary word ("race") with a more neutral one ("ancestry"). Elhaik's analysis was based on the knowledge that individuals in each group uniquely carried a particular pattern of DNA because they descended from a particular population of people. The concept of biological race is of no help here, not only because it is imprecise but also because it carries a fatally incorrect implication—the idea that people can be sorted into completely distinct genetic buckets. Ancestry does not work that way. Elhaik now works at the University of Sheffield· and the company Prosapia Genetics has been created based on his analyses.

Since the Human Genome Project popularized the incorrect idea that two individuals from different populations are often more alike than are individuals from the same population, people have tried to remedy the misconception of genetic race by portraying the human genome as a single continuum where all groups are like beads on a string. But that's not a useful metaphor either: You can't take Elhaik's hundreds of groups and place them in a single line. While the human genome can be described as a continuum, it is one that branches and changes through time. You can think of it as a tree with a definite and irreducible shape. The end of any healthy, growing branch is a population that exists today. The base of the trunk is the single population from which everyone alive today has emerged. The branches themselves may form a tangled thicket too, different twigs and branches often fusing together to form one.

If you took the genomic tree and made every part of it invisible but the ends of the branches, you would essentially have a map of current human populations. You would see that there were obvious clusters of people but also that there was continuity among the clusters. You might even be able to discern some of the world's geography in the population map; people are often most like the people who live near them. When scientists like Elhaik analyze the DNA of living populations, they effectively make the whole ancestral tree visible. By measuring ancestry in genomes, they reveal that we are both different—different groups emerge from different branches—and the same—we all emerge from and cluster tightly around the same trunk.

Is it dangerous to contemplate the tree? Despite the resistance to genetic information about history, there has been little research into how people actually use it. Though we are by now well educated about what we fear people will feel, we don't know much about what they actually do feel.

When Brian from Texas took a DNA test,* he had always believed that he was a mix of white, Cajun, and French Acadian. But he discovered that a significant amount of DNA on both sides of his family was usually seen in Native Americans. Although he had always identified as French, he no longer does. The results affected how he saw others too. Caucasians now looked different to him.

Brian participated in a survey conducted by Wendy Roth, the professor from the University of British Columbia who found her great-great-great-grandfather's name in a European cemetery. Roth was fascinated by how people's identities were affected by information about their DNA, and she felt that there was a "general level of ignorance, a lack of awareness, a lack of interest in this thing that should be worth studying."

She contacted DNA test takers and found that people's responses to news about their ancestry were often nuanced and complex and that they changed over time. Most did not experience a significant disruption in their sense of their identity, primarily because the data didn't contain any big surprises. When people got news they weren't expecting, for some, like Brian, it changed everything. Yet, Roth recalled, "There are very, very few people I spoke with who will completely change their identification." Generally the people who said they were changed by news about their ancestry *expanded* their sense of self to include the new information.

---

* All names in Roth's study are pseudonyms.

One man whose ancestry was Mexican American discovered that he had Celtic ancestry too. But he had little interest in talking about it because of the stereotype of Celts as physically large and the fact that he was small. People might think it was his fantasy. Others were quite happy to embrace the diversity and complications that their tests revealed, but they found that other members of their families were less open. One woman who considered herself black found that her genome was 39 percent European. While her response was curiosity, her sister did not embrace the news. Another woman who identified as white discovered African ancestry on her father's side of the family, so she started going to movies and plays exploring the black experience. Yet, she told Roth, she was unable to share the news with her bigoted brother.

Some respondents embraced the new information but became hesitant at the point where it threatened to change significant aspects of their lives. A woman who found out that she had Jewish ancestry was invited to the local synagogue. But the strictness of the religion and the new and different prejudices of the people she met made her feel that she did not belong.

While some felt positive about a newly discovered multiracial history, they were reluctant to announce it in case they were viewed as "wannabes." A number of Roth's test takers who discovered Native American ancestry felt the news was complicated by the availability of government money to that minority. They were afraid that people would assume they had produced these lost forebears as a way to access it. Others were concerned that they would be viewed as abandoning their "real" identity and trying to pass as something else.

Sometimes the lack of participation in DNA tests was telling too: Roth was unable to find many Asians for her study and suspects that Asians are less likely to take ancestry tests. When she asked Asians who had taken the tests about this, they would tell her: "Many of us think we know what we are." Roth observed that "such beliefs that their lineage is completely unmixed is likely no more accurate than for any other groups." She added, "There is the sense that Asians are very homogeneous in their roots. I think it is related to the national myths and stories that people tell about who they are."

Overall, Roth found that some people overinterpret DNA and some people don't, and some people react extremely but most do not; in short, Roth's responses run the usual human gamut, except for this one ray of light: When many people found out something new, their reflex was to increase their knowledge. One man who discovered that his mother's line was connected to the Fulani tribe in Africa began to learn the language.

"This kind of testing seems to make people more aware of how much racial mixing has gone on historically," Roth said. "I think a lot of people start out thinking of themselves as being 100 percent something, and they don't necessarily go into genealogy because they're trying to challenge that view, but as more people get immersed in genealogy and, especially, as they do tests like this, they realize that, no, they're not 100 percent something. . . . There is mixing that has happened, whether it's a long way back or whether it's just a couple of hundred years ago or whether it is within the last generation or two."

If we want to understand mixing, whether it's in our own family or in some larger group to which we belong, we have to understand DNA, but we also have to take into account its context. Jennifer Wagner is a lawyer and anthropologist who translates science for the legal world and vice versa. She advocates studying human differences "holistically, integrating the contributing factors of culture, sociology, history, genetics, evolutionary biology, and the like." With a group of colleagues she is working to develop an innovative curriculum to teach about evolution using genetic genealogy ("a more exciting . . . way of teaching these concepts than is the study of peas or fruit flies").

Currently in the United States, according to Wagner, "minorities are overrepresented in forensic databases and underrepresented in biomedical research databases. Genetic and genomic technologies could either mitigate or exacerbate racial disparities. We must be mindful of that and do everything we can to ensure that every individual shares in the benefits of scientific knowledge."

The big human family tree that Wagner will teach could have grown into thousands of different shapes, so why has it taken the particular shape it has? In part it is because of its biological machinery, and in part it is because of the events of history. Human choice, chance occurrence, and unpredictable contingency have all contributed to the tree's growth. It would be impossible to identify all the factors that have shaped the genome, but we are beginning to have the ability to piece together the events that matter. What were the biggest shapers of the genome we have today?

Chapter 12

The History of the World

Human beings are ultimately nothing but carriers—passageways—
for genes. They ride us into the ground like racehorses from generation to generation. Genes don't think about what constitutes good or
evil. They don't care whether we are happy or unhappy. We're just
means to an end for them. The only thing they think about is what is
most efficient for them.

—Haruki Murakami, *1Q84*

When you visualize the human tree, picture its trunk firmly
planted in African soil. Modern humans emerged there several
hundred thousand years ago and lived only there from 250,000 years
ago for at least 150,000 years—a much longer span of time than we
have lived across the globe.

Working out what life was like when humans were an exclusively African species is probably one of our biggest scientific challenges. There
are no written records and few fossils from that time, and handmade artifacts date back to only 70,000 years ago. Which is not to say that humans
didn't use tools or wear jewelry before then—it's simply that, if there are
any that remain, we haven't yet found them. Still, while scientists have
only begun to plumb these depths, with each year that passes our view
into the past reaches further back as we find new evidence. While we often think of human history as a kind of reverse dimming, in which the
light of our consciousness and intelligence grew ever brighter, the evidence that we were profoundly aware as early as 200,000 years ago is
growing. In the last few years 60,000-year-old ostrich eggs with marks
that appeared to be intentional engraving were discovered. Dating of
beads from Israel and Algeria suggests they are between 100,000 and

130,000 years old. Ancient tools found in Crete suggest that someone sailed there more than 100,000 years ago. It looks as if there was ocher processing in the Blombos Cave in South Africa 100,000 years ago, and in the same region people sharpened the tips of their stone tools using heat—a technique that we used to think dated back only 20,000 years. In the history of science few people have dared to imagine that humans were as intelligent or as technologically adept so early in their history.

For hundreds of thousands of years much of the African continent was inhabited by different family and tribal groups. Around sixty thousand years ago a small band of them—perhaps not much more than one thousand to two thousand five hundred individuals—went traveling. We don't know why they left or if they had any sense that they were going somewhere new, but we do know their decision kicked off one of the biggest events in the history of the human genome.

Those who remained are the ancestors of most of the one billion people who live in Africa today. The small band that left are the ancestors of everyone else in the world, and the suite of DNA they carried was only a small sample of the variety of human genomes that existed in Africa at the time of their departure. Indeed, we know the migration occurred because even now we can see that the genomes of everyone in the world outside of Africa is a subset of the genomic variation still found in Africa.

When a small sample of a species's genome is isolated and then becomes the foundation for another group, it's called a bottleneck. (In this analogy the neck of the bottle is the small founding group, and the expansion of the neck into the body is what happens when the population grows.) Bottlenecks can be caused by many things and are powerful examples of the role of chance in shaping the human genome. "Things that happened a long time ago can constrain what can happen afterwards," said Marcus Feldman, a professor of biological sciences at Stanford. "If there is a disaster that kills off 98 percent of the organisms of a certain type, then what happens to the rest of the animals is constrained by the fact that there are only 2 percent of this particular type that are left."

As a group passes through a bottleneck, it becomes particularly vulnerable to drift. DNA diffuses more quickly within a small population, and it may not take too many generations before everyone's genome starts to look a bit more like everyone else's. There's no rhyme or reason to drift; it's a matter of chance. Bits of DNA may spread throughout a group for no reason other than that the people who carried them ended up with more children and therefore passed them on more often. DNA that has drifted throughout a group may have consequences for the people in it—like red hair or a protruding brow or a certain health issue—or it may not. It is equally possible that some bits of DNA will not come into prominence but instead float off into oblivion.

The out-of-Africa bottleneck is one of the easiest to identify, but it's far from the only one in human history. Around seventy thousand years ago one bottleneck may have taken the human race perilously close to extinction. A volcano erupted in Toba, Indonesia, causing abrupt climate change and leaving a layer of ash over a huge segment of the world. Some researchers have proposed that all of us today descend from a small number of Toba survivors. Indeed, if you wanted to summarize the myriad migrations, cataclysms, illnesses, innovations, and acts of love and hate that have changed the human genome, you could say that it's been shaped by a series of bottlenecks, where a population shrinks, and fusions, where two or more populations come together and blend their genetic material. To be more precise, you'd have to throw in some Darwinian adaptation as well. Marcus Feldman, who has been comparing populations across the world for a long time, says that the differences between populations "are reflective of two processes. One is migration and the distance from Africa, and the other one, most of which happens after the origin of agriculture, is natural selection on some genes."

Although the first big out-of-Africa migration was enormously significant, as it marked our transition from being a regional animal to being a global one, there have been many significant migrations since then. Indeed, the histories of most of the world's large populations include a bottleneck of that first out-of-Africa population.

As they traveled, humans made their way through Asia and along the coast into the southeast. They passed through strange climates and terrains, stumbling upon fantastically colorful and unimagined wildlife, much of which tried to harm or eat them. Everywhere they stopped, they left descendants behind, and wherever the descendants stayed, they adapted to the local terrain and available food. Within a few generations these weird new worlds became a familiar landscape that the travelers' descendants had always known. In the course of time, later generations of their descendants ultimately changed, becoming different colors, shapes, and sizes.

The wanderers and their descendants invented the taming of animals like dogs, goats, sheep, cats, and horses. They invented transport like sailing and skating on ice. Some of the earlier groups of travelers met humanlike creatures, thickset survivors of an earlier exodus. Others who made it all the way to the landmass we think of as Indonesia may have discovered a group of people who were all the size of small children. On the first leg of the trip they traveled as far as possible, arriving in Australia, a land of two-ton wombats, ten-foot-tall kangaroos, and enormous marsupial lions, about fifty thousand years ago. The ancestors of modern humans began to spread through Europe only forty thousand years ago.

Less than 18,000 years ago humans arrived on a landmass now known as North America. (Only Antarctica has been free of humans for longer than the Americas.) It seems that almost the entire indigenous population of the Americas descends from a small group of perhaps eighty people, originally from Siberia, who followed a route that has now been covered over by the ocean. Over 32,000 years ago they took refuge in northwestern Beringia, a land bridge connecting what are now Alaska and Russia. Over the millennia that followed they moved into eastern Beringia, and then sometime before 14,000 years ago they moved into the North American continent, spreading along the Pacific coastline and then eastward. Geneticists have found that Native Americans have only five kinds of mtDNA, and the first four are common in northeastern Asia. With this evidence and other genetic studies, it has become well-established that the small ancestral population of Native Americans can trace its genome back to Asia.

The picture got more complicated when a study by David Reich found an ancient connection between the modern Native American and European genomes, suggesting that there once existed a population in northern Eurasia that was ancestral to both. The study was based on a comparison of modern genomes; no bones from such a population had been found. Yet in 2013 the remains of a young boy who lived 24,000 years ago were found in Mal'ta in south-central Siberia. Analysis of his DNA showed that he was related to modern Europeans and Native Americans. The finding confirmed that at least 14 percent and up to 38 percent of Native American DNA came from a population in western Eurasia. Remarkably, a few months later the scientists who analyzed the Mal'ta boy's DNA published the genome of another ancient boy, an infant who was buried in Montana more than 12,500 years ago. Anzick-1, as he was called, was covered in red ocher and placed in the earth with stone tools from the Clovis culture. He is the first ancient Native American whose genome has been sequenced. The people who buried Anzick-1 are ancestors of modern Native Americans (although he is more closely related to 44 groups from Central and South America than to others from North America).

The complex tales told by the ancient Siberian and Clovis children are echoed by what happened during the out-of-Africa exodus in the rest of the world. No matter where they stopped, even after a group of travelers settled in place, life—and the genome—kept changing. Some settlers were joined—or overrun—by others; sometimes a subgroup set off anew. It was thought that the Australian genome was isolated for tens of thousands of years before eighteenth-century colonization, but in 2013 it was discovered that roughly four thousand years ago a band from the Indian subcontinent traveled into Australia and contributed to the genome. Around the same time, there were changes in tools and in the way food was processed, and the dingo first appeared, suggesting that the Indian group may have brought the wild canine in with them.

Of course, it's not just the descendants of the out-of-Africa band who have changed; African populations have changed as well. Indeed, as far as the genome is concerned, the groups who remained were also small bands of travelers. There were genomic bottlenecks on the African continent long before the 60,000-year mark: In 2012 researchers announced that they had found one of the original branches of the human family tree. The Khoe-San, a tribe who live in southern Africa, split off from everyone else 100,000 years ago. In addition, many groups journeyed through the land and blended with other groups. In many parts of the continent native Africans effectively journeyed through different environments even as they stayed in the same place. In all that time the climate changed, plants bloomed, and animals thrived, and then ice ages dried the land out.

Feldman and colleagues have counted the number of bottlenecks that different modern populations have passed through and have found that populations that have passed through the most bottlenecks have more deleterious mutations in their genome than populations that have passed through fewer. Yet even as they find ways to identify difference, their work still underlines the overwhelming commonality of all people. When you examine the human genome, Feldman told me, "The thing that strikes you is that people in different continents actually have very similar genomes and that the fraction of the genomes that are different is pretty small. I mean, you're down to a tenth of a percent."

When humans left Africa 60,000 years ago, it was almost certainly not the first journey they had attempted but merely the most successful one. The bones of modern-looking humans found in the Skhul and Qafzeh caves in Israel date to 120,000 years ago. These people are not our direct ancestors but were likely an earlier group who walked out of Africa. It may be the case that the out-of-Africa journey that led to the peopling of the world was more complicated too. A 2014 study that compared both DNA and the shapes of fossilized human skulls suggests that the ancestors of the Australian Aboriginal population actually left Africa 130,000 years ago and that there were at least two waves of the modern human exodus from Africa. Stone tools found in the inland deserts and mountains of the Arabian peninsula that date to more than 100,000 years ago support this idea. Yet another kind of creature left Africa much earlier, almost 500,000 years ago, and founded a civilization that spread across much of the world.

There's another way we can tell that the human tree splits between people who come from Africa and people born elsewhere in the world.

The traces of the split lead us back to an event that occurred just as the out-of-Africa diaspora began. When the small band of travelers was essentially standing on the doorstep of the continent, perhaps wondering where to turn, they met up with a group of Neanderthals and ended up making some human-Neanderthal babies together. All non-Africans today carry the mark of those encounters in their DNA.

In just the last few years we have learned that 85 percent of all people carry DNA from Neanderthals, an entirely different species that lived until 27,000 years ago. If the research on the human genome hasn't completely destroyed the idea of genetic purity, our newly discovered Neanderthal ancestors show how truly absurd the notion is. Colin Groves, a professor of bioanthropology at the Australian National University, explained, "Neanderthals and *Homo sapiens* are like lions and tigers. Genetically they are sharply distinct, but they can interbreed."

A first-draft sequence of the Neanderthal genome was published in 2010 by an international team of scientists, including David Reich at Harvard. I visited Reich's lab in 2011 and asked him what that first meeting between our ancestors was like. He explained that it might have been a meeting of a few dozen humans and Neanderthals, or it could have been a blending of thousands of individuals. At the time I spoke with Reich, we didn't know which parts of the human genome had come to us from Neanderthals. Since then the science of Neanderthal DNA has progressed faster than anyone imagined it would.

The boom in Neanderthal knowledge comes from a revolution in the science of ancient DNA, much of it led by Svante Pääbo from the Max Planck Institute in Leipzig, Germany. Ancient DNA is the most difficult kind of DNA to study. For a long time it was thought that DNA could not survive beyond days or weeks, yet scientists can now locate and extract DNA from fossils that could be tens of thousands of years old. The first Neanderthal genome was extracted from bones that were found in a cave in Croatia dating to more than 38,000 years ago. The technical challenges of reading ancient DNA are so complicated that it was thought to be simply impossible, a situation that is further complicated by the enormous risk of contamination from modern human DNA. Only a few labs around the world have built sterile "aDNA" labs to protect against this.

Now, in addition to all the different ways of finding out about the past through the genomes of modern humans, we have begun to build a library of ancient genomes. We can compare the DNA of ancient people to modern humans and also compare the ancients to one another. Reich led a pioneering study that compared the mtDNA of 364 ancient individuals who lived in one of nine different European cultures between 1,550 and 5,500 years ago. (Because there are so many copies of mtDNA in any one cell, it is much easier to recover from ancient remains than nuclear DNA.) The team found that the way mtDNA transformed over time revealed a pattern of stasis interrupted by change. After farming was first introduced to central Europe, not much about the genetics changed for 2,500 years; after that the genetics associated with the farmers began to spread. Reich's team found four significant incidents where a population either expanded or was replaced by another population which they lined up against significant cultural moments, such as the introduction of the horse and the beginning of metallurgy.

A 2014 study used the DNA of ancient farmers and hunter-gatherers from Europe to explore an age-old conundrum: Did farming sweep across Europe and become adopted by the resident hunter-gatherers, or did farmers sweep across the continent and replace the hunter-gatherers? The study found a significant difference between the DNA of the two groups, suggesting that even though there may have been some flow of hunter-gatherer DNA into the farmers' gene pool, for the most part the farmers replaced the hunter-gatherers.

Now we have answers to the questions that couldn't be addressed just two years ago. What exactly is Neanderthal DNA doing in the human genome? Is it merely the reminder of a long-ago encounter, a random series of segments that have drifted throughout the genome? Or were some segments of Neanderthal DNA retained because they shaped us in useful ways?

Even though most non-African individuals have 1 percent to 3 percent Neanderthal DNA (I have 2.7 percent), it appears that over 60 percent of the Neanderthal genome is distributed in small pieces throughout the non-African human population. A number of teams have demonstrated that it may have helped the earliest African migrants adapt to a colder, darker climate. Some parts of the genome with a high frequency of Neanderthal variants shape hair and skin color and likely made the first Eurasians lighter-skinned than their African ancestors. Other regions that have been influenced by the Neanderthal genome are implicated in human diseases, such as lupus, Crohn's disease, and type 2 diabetes, and even in behavior, such as addiction to cigarettes. Some Neanderthal DNA appears to be more useful to one population than to another. Europeans, but not East Asians, have more Neanderthal DNA in regions of the genome responsible for lipid catabolism, the processing of cholesterol and fatty acids and related molecules.

It may even be that some Neanderthal DNA was selected against. There are regions in the human genome where no Neanderthal DNA can be found at all, such as genes that are significantly expressed in human testes. Perhaps the individuals who first inherited those segments were not successful at passing them on.

Not long after the Neanderthal genome was sequenced, Pääbo's team discovered that some people carry DNA from an entirely different ancient species, now known as the Denisovans. Until 2010 we didn't even know Denisovans existed, and although all we have of them today is a few tiny bones and a couple of teeth found in a cave in the Altai Mountains in Siberia, scientists were able to extract DNA from these remains and compare it to the genomes of modern people. The Denisovans may have spread as far as Southeast Asia. Indigenous Australians, Melanesians, and some groups in Asia have up to 5 percent Denisovan DNA, in addition to their Neanderthal DNA. It's been suggested that an early group of humans may have left Africa, met with Denisovans in Asia, and then spread their genes out from there, bringing them into Australia more than fifty thousand years ago.

The only group that doesn't seem to have traces of either Neanderthal or Denisovan DNA—at least from this period of history—is sub-Saharan Africans. Yet researchers are examining the African genome for evidence of earlier mixing with archaic human beings. In 2011 it was announced that some Africans carry DNA from an entirely different, as yet unknown, species.

The fact that we carry this ancient nonhuman DNA changes how we view not only humans and nonhumans but also the entire narrative arc of ancient history. We've always imagined the migrations of humans out of Africa as a hero's journey, with a small band of gutsy wanderers setting off intrepidly into the unknown. But it seems that even then there was no terra nullius. The residents of planet Earth included not just Neanderthals and Denisovans but at least one more mysterious population: the hobbit, a tiny humanlike species who lived as recently as thirteen thousand years ago on an Indonesian island.

While the mixing of Neanderthals and humans was a special case of admixture because they were so distantly related, there have been many significant examples of populations merging in history. "The greatest changer of genetics in history—probably," Marcus Feldman said, "has been colonialism, whether it's the Mongolian invasion of all of central Asia, which left genes lying around all over the place, or the British colonization of Australia, which left a very large signature of British genes in the Aboriginal population, or the Hispanic colonization of the Americas."

As much as populations have split apart and bottlenecked, there have also been continuous waves of humanity, sweeping over one another and fusing, whether it's a merging of two separate populations or the subsuming of one, where only a very small trace of either the invaders or the invaded remains in the genome.

Genomics allows us to see effects of colonialism that took place thousands of years ago. Feldman and his group showed that beginning around five thousand years ago, the Bantu began spreading throughout Africa, and by three thousand years ago they had reached southern Africa, where they began to merge with many local groups. Because they were farmers, the Bantu pushed out many of the native populations, who were hunter-gatherers. Descendants of Pygmies who developed a working relationship with the Bantu now carry significant evidence of Bantu ancestry in their genomes. By contrast, said Feldman, Bushmen from

Namibia, who haven't been much in contact with either people of European ancestry or people of non-Bushman African ancestry, "don't show much of any ancestry other than their own."

It's possible to see how old the Bushman genome is by looking at how tightly Bushmen's genes travel together when they are inherited. "Think about beads on a string," Feldman explained, with each segment of DNA representing a bead. "Every time there is a generation, two beads next to each other have a chance of breaking and forming a new string. If that happens at a certain rate, then the likelihood that you'll find your original two beads on the same string after a long time is pretty small." This is linkage disequilibrium, and Feldman and his group have found that Bushmen have the lowest amount of linkage disequilibrium in the world, meaning their genomes have been cycling over and over for the longest amount of time.

For most of the human genome's history, before the age of mass transport, bottlenecks and admixtures have occurred at the speed of walking. But in more recent history the geographic trail has significantly decoupled from the genetic trail, as every mode of transport we have invented has hastened the splitting and mingling of the genome today. As Marcus Feldman pointed out, it became possible to carry out colonization on a massive scale in the era of the horse, as we see in Genghis Khan's legacy.

Still, it wasn't until the development of big ships and the age of exploration, the slave trade, and large-scale immigration that huge genomic waves began to wash around the world. In 1511, when Portuguese apothecary Tomé Pires rode the trade winds to Malacca (a state in modern-day Malaysia), he discovered a multicultural port city where he counted over eighty tongues being spoken, including ones from Europe, Africa, Eurasia, China, and the islands of the South Pacific.

By the time of Columbus's voyage, slave trade among Europe, Asia, and the Middle East had been a key component of commerce for hundreds of years. Following the discovery and settlement of the Americas, including the Caribbean and Brazil, newly established sugar and cotton plantations required a massive influx of labor, which was supplied by slaves from West Africa. Today Brazil is second only to Nigeria in terms of population with ancestry dating back to the African population of the

Middle Ages. Overall, 5.5 million Africans were shipped to Brazil between 1501 and 1866.

Even before the potato famine of the mid-1800s, the Irish had begun leaving for the United States, Canada, and Australia. During and following the famine, emigration increased dramatically, and by 1890 40 percent of all Irish-born people were living outside of Ireland. Today, there are seventy million people worldwide who claim Irish heritage, of whom only five million live in Ireland. Before these modern emigrations, much of the Irish population had inhabited the island we call Ireland for thousands of years.

The genome not only is a record of the fact of admixture but it can tell us something about how the mixing occurred. "The Y chromosomes of indigenous Americans is heavily biased towards European, whereas the mitochondria are not," said Feldman, explaining that this reflects the fact that the colonists were all men. As they swept in, they killed much of the male population, effectively removing their Y chromosome, and bred with the indigenous women, whose offspring inherited the colonists' Y. The children retained the mtDNA that was passed down by their mothers.

This pattern is true for many populations, including the African American population. You can see in the modern genome that stories like that of Jefferson and Hemings were not an exception. Many African women bore the children of white men.

According to Nick Patterson at the Broad Institute, when you have multiple waves of male invaders washing over a female group, it's not just the Y that changes. While the mtDNA stays the same, the autosomal DNA may be replaced completely. You can also track more complicated patterns of history in the X chromosome because two thirds of the ancestry on the X chromosome is female (women have two X chromosomes and men have one). The signal of the X is more complicated than that of mtDNA, as it is a combination of male and female, but it's weighted toward female history.

When I think about the forces that have changed our genome, I imagine massive apocalyptic clouds rolling across a dark sky, a world covered in ice, or a famine that leaves only the luckiest few standing. Or I just picture time, hundreds of decades rolling on, one after the other, crunching up and spitting out generations of people behind them, continually transforming the genome. But natural selection is not all cataclysm. Some events remolded half the species without any such thunder.

Imagine a drink that, if consumed regularly, would completely shape the lives of the drinker's descendants, generation upon generation, over many tens of thousands of years. It sounds like science fiction, but it's ancient history. Before eight thousand years ago, humans could not easily digest milk after they had been weaned from their mothers. As people learned how to domesticate first goats, then sheep and cattle, they tried drinking the animals' milk and then came to rely more and more on it for sustenance. In many parts of the world this became a crucial determinant of who was most likely to live and reproduce and who was not. Random mutation meant that some people could tolerate milk better than others, and these individuals passed on their tolerance to offspring who outsurvived the lactose intolerant. Milk drinking evolved several times over in different groups.

The milk adaptation comes from the biology of culture, where ongoing changes in the human body resulted from choices people made as they created their environment. The way we normally think about natural selection (aka adaptation, aka survival of the fittest) is that a child is born with a genetic mutation that gives it an edge—whether stronger immunity or greater height—that enables it to be more successful in reproducing than its peers. Because the edgier offspring pass on their advantageous trait to more offspring, the new trait—and the DNA that underlies it—becomes more frequent in the population and possibly completely dominant. When natural selection shaped us in Africa we were a small enough group that it shaped the entire human genome. The gene for amylase is an example of the biology of culture via the ancient human kitchen. Amylase helps people process starch, and it was discovered in 2007 that the more starch a group of people eats, the more copies of the amylase gene they have. It's not known if the starch-eaters have gained copies over time or if the non-starch eaters have lost them.

It may be that one of the threads of the human journey is underscored by amylase—the better we could process starch, the more widely we were able to travel.

As time went on and different populations moved around and split into different groups, natural selection impacted different populations in different ways because it was also shaped by their different behaviors. According to Marcus Feldman, most of these changes were in response to the transition from a hunter-gatherer lifestyle to an agricultural one: "What we're talking about here are the genes that allow you to use milk, genes that allow you to use wheat, which was not in the diet of the preagricultural people. Those kinds of things were stimulated by farming ten thousand years before now."

In the course of evolution humans have also lost huge numbers of working genes along with the abilities and traits they generated. In recent years scientists announced that human bitter-taste-receptor genes are losing their function. Identifying bitterness helps animals avoid toxic foods, precisely because so many of them taste bitter. For humans, however, an increase in meat consumption and a decrease in plant food, as well as the use of fire, which renders many toxins harmless, have meant that these bitter-taste-receptor genes are no longer maintained by natural selection. As a result they have effectively become useless.

The degradation of our taste is only one element in a much more massive loss of functionality. In comparison to our ancient selves we now have a greatly diminished sensitivity to many diverse signals from the natural world. Lots of human genes related to smelling, seeing, and the ability to identify pheromones no longer work. Humans have four hundred functional olfactory receptor genes left from what was once a much larger set. Mice, by contrast, have more than one thousand functional olfactory receptor genes. The genome of modern mice is thus much more like that of the distant human and mouse ancestors of millions of years ago. In addition to smell and taste, it's been suggested that even the range of human hearing is narrowing. In a few million years will we be deaf, taste-blind, and unable to smell anything? It's not the direction we usually imagine that evolution will take us.

Four hundred and sixty years ago, at the beginning of the age of travel, a series of apocalyptic Mexican pandemics killed tens of millions of indigenous people, who died from a previously unknown (and never-again-seen) hemorrhagic fever. One pandemic killed 80 percent of the native Mexican population. The next one killed 50 percent of the remaining population. Coupled with devastating droughts, the introduction of smallpox, and terrible treatment at the hands of colonial invaders, the indigenous population of Mexico was devastated and its genomic diversity was forever reshaped.

You might expect that the events of history that have most deeply shaped our psyches are the same events that have carved the genome. In some cases this is true. For the hundreds of years that the slave trade operated, it changed the lives of millions of people, and it may have left a legacy of individual distrust, as well as economic dysfunction, for generations after it ended. The slave trade also completely altered the topography of the human genome. The modern genome of many populations in countries into which slaves were transported reveals descent from them but also from the slaver population and other migrants.

And yet here is the odd reality of the human genome: The history of the world may be writ in your cells, all of it personal to your lineage and some of it part of the broader context, but though you have been shaped by history, you have only been shaped by some of it.

Fundamentally, disease and other catastrophes that give rise to bottlenecks affect the genome of future generations by bottlenecking the ancestral gene pool. Similarly, an event that wipes out a whole subpopulation is a genomic event because it removes a variety of the genome from the overall library. In that context landmark events like World War I or World War II or the 1918 flu pandemic may not be genomic events in the same way, even though, for example, the flu killed people from populations all over the world. While the virus profoundly affected some relatively small populations, such as the indigenous residents of Western Samoa, which had proportionately higher fatalities, they recovered. Western Samoa is now a self-sustaining population.

The events of history that left a trace in the genome are not necessarily history's most important events either. There could be many significant events that didn't leave a mark, which is why ancient DNA is such an important tool. We can't find ancient individuals whose Y chromosomes no longer exist today by examining the modern genome.

If lice had names, you could, theoretically, create a genealogy that connected all the lice that had ever lived to all the people whose heads had provided a home for them and then trace the joint genealogy of lice and men all the way back to Africa. In lieu of such records, the newly sequenced genome of the louse has confirmed the broad outline of the story. By identifying the four major subpopulations of lice, researchers have shown that their genomic tree can be overlaid on the human tree, and the story of both species' migration can be traced together. A similar tale can be told in the Pacific by the divergence between two kinds of *Helicobacter pylori*, the bacteria that can cause stomach ulcers. *H. pylori* originated in Africa with humanity and it too has been shaped by subsequent bottlenecks and isolation. Researchers found that an ancient migration of people into New Guinea and Australia brought along an ancient version of the bacteria. By contrast, the relatively recent migration of a different group into Melanesia and then to Polynesia is validated by the way the DNA of their *H. pylori* has diverged from the ancestor of the New Guinea/Australia bacteria.

As we learn to interpret the marks that the world has left on our own genome, we also acquire the ability to see how the world has shaped others species' genomes. Curiously enough, when we play a significant role in an animal's world, we can see our history in its genome too. The domestication of farm animals has completely changed their biology. At first the sculpting was somewhat accidental, but since Robert Bakewell formalized the enterprise in the mideighteenth century, we have been intentionally selecting and breeding cattle, horses, sheep, and other farm animals. The domestication of dogs and cats has been generally less utilitarian but no less significant for their genome.

In 2008 geneticists at the University of York discovered that mice have left genetic trails in much the same way as humans. Rodents that traveled into Orkney on Viking ships ended up leaving much of their

DNA in the mouse populations on the island. Indeed, the Scandinavian mice left a pattern so clear that scientists have found they can draw an accurate map of human movements based on mouse movements alone. A more recent study tracked marauding mice of the early tenth century into Greenland from Iceland and before that from either Norway or the northern part of Britain. The researchers looked for, but did not find, Viking mouse DNA in the mice of Newfoundland. Like the short-lived New World Viking settlement, the Viking mice did not leave a lasting imprint.

Similarly, certain ancient maritime trade routes and the spread of pastoralism from Africa throughout the Near East and the Mediterranean coastline are revealed by an analysis of the Y chromosome and mtDNA of modern-day goats.

In the same way that looking back into our immediate family's past may change how we think about time and history and our place in it, so too does taking on the idea of our more distant ancestry. Once upon a time, history was living memory plus all the increasingly fuzzy spans of time that came before it. Now we may use written records and the artifacts and fossils that came before records. Using all of these sources of information with DNA teaches us simultaneously about human history, the forces of evolution, and ourselves. Ancestry brings together history and science without any artificial seams between them. It explains our immediate family in the context of the human family and vice versa.

Ancestry takes account of the evolutionary mechanisms that have shaped the human species *and* the one-off events—the drifting of genetic markers and traits that have also shaped different parts of the genome. It includes all of human history, from our great-to-the-200th-degree-grandfathers to our two most recent grandfathers and the social and genetic legacies that they passed on. It implicitly includes each of us, because the end result of any lineage is—at least at the present moment—after all, us.

Ancestry teaches us that when people spend long periods of time together, they form populations. It opens the door to the idea that those populations are shaped by both drift and selection. It shows us that members of a population may share physical traits, but they also share genetic markers of their history that don't have anything to do with their traits. Ancestry teaches us that we can learn more about our children by knowing their grandparents.

The study of ancestry, of course, has its complications. Examining the threads of heredity and influence through time can threaten what we tell ourselves about self-determination, both on the social scale and on the personal scale. This happened when people in the United States turned to caring more about their American heritage than their British, and this is surely true for other postcolonial countries. But it applies to our personal lives too.

Genealogy doesn't always lead to an enlightened sense of self either. When describing the heritage of the English upper classes, one English artist spoke of the objects in their homes as having been chosen only because they signaled "historic affiliations." He saw their family trees projecting from their foreheads "like antlers," weighing them down with responsibility.

Still, if you find your ancestry explains something about your experience of yourself or your family members or your body, then it's a worthwhile pursuit. If your ancestry provides you with a framework for a cluster of ideas or thoughts or feelings that you have never connected before but that suddenly seem related in light of what you've learned, then it's not only interesting but also productive.

As for the ever-controversial subject of race, if you find it's too hard to let go of the word, it may help to redefine it as *ancestry*. In this sense, race is a record: The traces of "race" in our genome are what is left of our ancestors' lives. In the same way that the medical community is beginning to explore the idea of personalized medicine, you may find that thinking about personalized race helps you better utilize information about individual and group differences when you consider your history or your health.

In medieval times monks wrote on animal skins, and when they needed more vellum they scraped the original writing from old texts and wrote over it. With the help of modern technology it is often possible to make out not just the monks' final words but also the earlier ones because of the traces they left in the vellum. In addition to the story told directly by the words, the way they were layered over one another also tells a story. These documents are called palimpsests, and now scientists have learned how to read another layer in them by analyzing the DNA of the vellum so as to learn when the animal whose skin it once was lived.

Our DNA is a palimpsest too, and thousands of stories have left their traces in our personal genome. As humanity evolved and traveled, and as families do the same, new stories are layered over old ones, and we can learn more by understanding when and how they were written over one another. (Cultures are palimpsests too. The way we make decisions, such as whom to trust or whether to get divorced, may lead us back to personal events or much larger social events from a long time ago.)

But DNA is not only a record of history; it is also the stuff of which we are all made—an evolving set of instructions for the construction and operation of our bodies. DNA can shape how we feel, how we behave, and what we look like, and, of course, all of these qualities can shape how people treat us.

DNA and our life experiences make our bodies palimpsests. As we learn how to interpret the body in the context of its genetic code, we begin to understand how the hand of fate, the choices of families, and the enormous journey of DNA through deep time affect our lives right now.

Part III

How What Is Passed Down Shapes Bodies and Minds

Chapter 13

The Past Is Written on Your Face: DNA, Traits, and What We Make of Them

It is hard to hide our genes completely. However devoted someone may be to the privacy of his genotype, others with enough curiosity and knowledge can draw conclusions from the phenotype he presents and from the traits of his relatives.

—Philip Kitcher, *The Lives to Come:*
The Genetic Revolution and Human Possibilities

Wayne Winkler's parents moved to Detroit in the 1950s, but each summer when Winkler was a child they would return to Hancock County, Tennessee, to visit his father's family. On one trip when he was twelve years old, Winkler read an article in the Hancock County *Post* about Melungeons. "One of the most fascinating mysteries in Tennessee lore concerns the unknown origins of the Melungeons," it began. The article went on to describe the Melungeons as "a dark-skinned people whom some romantics compare in appearance to Othello immortalized by Shakespeare."

Winkler wanted to see some of these elusive Melungeons. He went to ask his father about them, but the elder Winkler had little to say on the subject. Later Winkler's mother told him that, in fact, his paternal grandmother was a Melungeon—which meant that not only was Winkler's father a Melungeon but Winkler was too. He later discovered that his paternal grandfather was a Melungeon as well.

Although Winkler had always known that his family had Native American and white ancestry, this new affiliation was a complete surprise. "I had always assumed that my dad's family was mostly Indian, because that's what they looked like and that's what they always said," Winkler recalled. When he eventually asked his father why he had always described himself as being an Indian, his father replied, "Every-body knows what an Indian is. It takes all day to explain what a Melun-geon is."

Despite—or perhaps because of—the fact that part of his family history had been obscured, Winkler was enthralled, and he felt proud to be part of a new group. The timing was right too: Winkler learned about his Melungeon ancestors the same year that James Brown's "Say It Loud, I'm Black and I'm Proud" was a hit. As Winkler described it, being born at the beginning of the civil-rights movement meant that even though he grew up in an era in which the South was still segregated, there was a growing consciousness and pride among many American groups who didn't pass as white. In 1968 a local Tennessee group even began to put together a play about Melungeon history.

Winkler felt encouraged by the liberal tendencies of the times, and the fact that his father and aunt were open to his questions. "My aunt Hazel gave me a collection of articles she'd copied from old newspapers and magazines," he explained. "That was my first research material. But it didn't answer questions; it raised questions." But other members of Winkler's family were not open, and it did not become any easier to find out more about himself or his ancestry. When he asked some relatives about being Melungeon, Winkler recalled, "They would say, 'Yeah, I don't remember that stuff.' It was just kind of putting me off."

Winkler had one uncle who offered little but still managed to com-municate a great deal. "He was kind of a character," Winkler recalled, "kind of a volatile character, particularly when drinking. He was drink-ing quite often, so he was somebody you kind of watched yourself around." Still, the old man was indulgent toward Winkler to an extent other family members found surprising. Years after the uncle's death, Winkler's mother said to him, "Your uncle would never have talked about being Melungeon." But Winkler replied, "Mom, let me show you something," and handed her a book. It was Jean Patterson Bible's *Me-lungeons Yesterday and Today*, published in 1975. Winkler's uncle had given him an inscribed copy as a gift. "I think that was his way of say-ing, "Yeah, okay, this is who we are," said Winkler.

When Winkler was a boy, the same uncle had taken him for a walk near the family farm. "He showed me this little unmarked graveyard, and he said—I hate to use this word, but he said—'What this is, is a nigger graveyard.' There were just a few headstones in it, and I said, 'That's our names on there. My grandmother Stanley's name is on there.' And he said, 'Yeah.' That's all he said."

Who were the original Melungeons? Where did they come from? Finding the answer became a lifelong interest of Winkler's, and he spent years researching the topic. At the most basic level the Melungeons, so called by the people around them and then increasingly by themselves, were a group of interconnected families who lived in certain counties in Tennessee, Virginia, and Kentucky. Many of the same surnames occur again and again in Melungeon groups: Bunch, Goins, Collins, Kennedy, Miner, Mullins, Osborn, Bowman, Moore, and Wright. The name on Winkler's grandmother's headstone was Givens, her maiden name. Another headstone in the plot was engraved Bunch.

Mystery surrounded these people for hundreds of years. Old photos of Melungeons from the last two centuries are a study in the unusual. According to various historical documents and oral histories, their features were Caucasian but they had dark skin, dark eyes, and thick, dark hair. The Melungeons were considered nonwhite by the white people around them, but no one knew what or who they were. Some believed they were a mix of white and Native American. Others thought they were the progeny of white people and escaped slaves. Others believed they descended from all three. In Winkler's childhood "Melungeon" was a term of abuse. Adults who grew up in neighboring white hamlets remember being told to behave, or the Melungeons would come get them.

Melungeons, along with other nonwhite groups in the United States, experienced many episodes of legally sanctioned discrimination. Ever since Europeans arrived on the North American continent, people of different races had been subject to different laws. According to Winkler, in colonial times the British outlawed marriage between two different races: In 1662 they banned marriage between blacks and whites; in 1691, between whites and Indians. In 1846 eight Melungeon men were prosecuted for voting, which was deemed illegal for them "by reason of color." In

1924, Virginia's Racial Integrity Act increased the penalties for interracial marriage (which was already illegal). It also redefined who was and who was not "white," so that marriages which might have been legal before, for example, between Indians and whites, were now prohibited. Virginia banned marriage between whites and nonwhites.

In Virginia and Tennessee especially, public antipathy toward Melungeons was unrestrained. In 1890 a Tennessee legislator said: "A Melungeon isn't a nigger, and he isn't an Indian, and he isn't a white man, God only knows what he is." He went on to say, "I should call him a Democrat, only he always votes the Republican ticket." Another senator described his rival to a journalist as "tricky as a Melungeon." He elaborated that a Melungeon was a "dirty sneaky thief."

Yet there was always curiosity about them too. As early as the late nineteenth century, newspapers published articles about the mysterious Melungeons. In one famous series published in 1890, written by a woman named Will Allen Dromgoole (who spoke to the senators described above), Melungeons were described as poor and despised and in the habit of distilling spirits. Dromgoole's series was a uniquely valuable and much-quoted record of Melungeon lives, but from today's vantage point it is difficult to distinguish between her descriptions of the extreme poverty, social isolation, and stigma experienced by her subjects and the stigmatizing that the journalist was herself doing. She wrote:

> They look for [the train] constantly . . . bringing joy to the cabin even of the outcast and ostracized; ostracized indeed. Only the negroes, who have themselves felt the lash of ostracism, open their doors to the Malungeons.

> They are exceedingly lazy. They live from hand to mouth and in hovels too filthy for any human being. They do not cultivate the soil at all. . . . They all drink, men, women and children. . . .

> After the breaking out of the war, some few enlisted in the army, but the greater number remained with their stills, to pillage and plunder among the helpless women and children. Their mountains

became a terror to travelers; and not until within the last half decade
has it been regarded as safe to cross Malungeon territory.

Today Melungeons are thought to be triracial. Indeed, there are many
groups who have lived in the United States who, since records were first
kept, have been called either one of the "little races" of the South or a
lost triracial group. Like Winkler's clan, they are large, related family
groups who have been long identified by their local communities as "not
white."

It's thought there were at least two hundred triracial groups in colonial
America, including the Guineas of West Virginia; the Croatan of North Carolina, South Carolina, and Maryland; and the Wesorts of Maryland. As with
the Melungeons, their origins are unknown, and their name was often used
as a term of derision. The Issues of Virginia were so called because the term
"free issues" had been used to denigrate "free blacks" before the Civil War.
The name "Wesorts" allegedly came from the phrase "we sorts of people"
(as opposed to "you sorts"). The Guineas were named after the English coin
used in the United States during the Revolutionary War.

These groups often acknowledged Indian heritage in some form, although historically they didn't live with known tribes in a traditional
culture. Now some have been recognized as Native American by federal
and state governments. In 2011 the Wesorts were recognized formally by
the State of Maryland as descended from the Native American Piscat-
away. Although many groups have long denied any African American
ancestry, it's thought that this was primarily because of the prejudice
they would encounter. Now it's widely believed that the ancestry of the
Melungeons and other "little races" includes African Americans—as
indeed does the ancestry of many modern white Americans.

But other origin stories exist for these groups as well. One legend of
the Melungeons had it that their forebears were survivors from the lost
colony of Roanoke, an attempt at settlement in 1584 that failed so completely that not a soul was found when the British went looking for
survivors. Some suggested they were shipwrecked pirates. More dramatically it was proposed that they were ancient Phoenicians who had sailed
to the New World in ancient times and remained, mingling with the

natives. Various historical documents record that Melungeons identified themselves as Portuguese—or, as they said, "Portyghee." Another legend has it they are at least part Turkish in origin.

N. Brent Kennedy, who wrote *The Melungeons: The Resurrection of a Proud People* and whose personal journey was similar to Wayne Winkler's, argued that Melungeons descend from a group of Mediterranean sailors who landed on the North American continent decades before the establishment of Jamestown, the first continuous European settlement. Kennedy believed that the Melungeons were a group of stragglers abandoned on the continent by Juan Pardo, a Portuguese sailor who was employed by the Spanish. Pardo built a fort and, according to Kennedy, left behind a diverse collection of fellow Portuguese, Moorish, French Huguenot, Turkish, and Iberian prisoners. Kennedy believes the group traveled inland and ended up settling into their own community with natives from Virginia and North and South Carolina. The Turkish and Moorish contribution to this new population was strongest at this time.

One of the arguments Kennedy uses to support his theory is the uncanny similarity of words in the local Indian dialects to some words from Turkish. According to his research, "Tennessee" is like *tenasuh*, which means "a place where souls move about." "Kentucky," which is Indian for "dark and bloody ground," is like the Turkish *kan tok*, which means "saturated with blood."

Even the word "Melungeon" has many fascinating origin stories. "Melungeon," say some, derives from the French *mélange* (mixture), from when a French colony lived near the Melungeon settlement in the eighteenth century. Others have suggested that it derives from the Afro-Portuguese *melungo* (shipmate) or the Arabic *melun jinn* (cursed soul) or the Turkish *melun can* (one who has been abandoned by God). The most romantic candidate is the old English term *malengin*, meaning "evil machination; guile; deceit." The word appears in *The Faerie Queene*, written by Edmund Spenser in 1590:

> So smooth of tongue, and subtile in his tale,
> That could deceive one looking in his face;
> Therefore by name Malengin they him call.

The Faerie Queene was well known in America's first European colonies. In addition it is said that Melungeons used archaic words long after they had been discontinued in other English-speaking populations.

Clues to the origins of the Melungeons may also lie in a cluster of physical traits said to recur in the group's families today. These include the grandly named Anatolian bump, which is described by many as an unusually large protrusion on the back of the head. Sometimes described as doughnut shaped, it is an exaggerated protrusion at the point where the skull turns from one angle to another. So-called shovel teeth—an indented hollow at the back of each incisor—are reportedly another common trait in the Melungeon group, as is the palatal torus, a bony protrusion at the top of the palate.

Brent Kennedy believed that certain diseases were prevalent among the Melungeons, including sarcoidosis, thalassemia, and familial Mediterranean fever. Kennedy's own investigation of his family's lineage began when he was diagnosed with sarcoidosis.

It is a strange experience to discover that you are a member of an almost mythical group. When Wayne Winkler first met the woman who would become his wife and told her he was Melungeon, she reacted, wrote Winkler, as if he had announced he was a leprechaun.

"She thought that's what the Melungeons were, that it was just a story, a folktale," Winkler recalled. "There are all sorts of folk legends around about the Melungeons. As a matter of fact, one day the History Channel sent a crew to Kingsport, Tennessee, and in a parking lot of a shopping center they asked people at random, 'What do you know about the Melungeons?' People said they were giants, they were cannibals, they lived in trees, the wildest things. Things that I had never heard."

Winkler published a powerful history of Melungeons, *Walking Toward the Sunset*. His Aunt Hazel and his father had been the most willing to share their pasts, and although his father was dead by the time he wrote his book, his aunt was proud of him. "They were the youngest children in their family," said Winkler. "I think they suffered less discrimination than the older kids." Still, during the writing of the book he had to deal with people who thought that Melungeons were a fairy tale.

But after the book appeared, Winkler said, "There were quite a lot of people that were unhappy about my talking about the family like that, even though I barely mentioned my own personal connection."

It wasn't just the two contradictory beliefs—that Melungeons don't exist *and* that they do exist but no one should admit it—that made research so difficult for Winkler. Melungeon history is enormously complicated, mostly unwritten, and in many respects remains hidden. Exactly who was a Melungeon was never fully recorded or formalized; everyone simply knew who others in the group were. Now, because the people who lived through those times are dead, all of that social complexity is lost. "Anybody who is involved in any sort of family research, they all find out that nearly everyone who could give them good information has passed away by the time they thought to ask the questions," Winkler explained. "You're always a little bit too late to get the good answers, to feel the thoughts of the people who might really have been able to tell you something."

Of his reluctant relatives Winkler observed, "I think they had a sense of shame that they weren't considered good enough, but the way the discrimination happened was really strange. It wasn't formal." When his relatives were young, said Winkler, authorities "just said, 'This is the school you go to' and 'Here's the school these people go to,' and everybody just kind of knew why."

Even when social attitudes became more liberal, there was little clear acknowledgment about what life had been like and why it was changing. "That all just sort of disappeared in a way that I haven't really been able to put my finger on," Winkler said. "I've talked to people who were around then, and nobody seems to know what exactly happened. But right around the time of World War II, the separation between Melungeons and non-Melungeons just kind of stopped. They started identifying Melungeon men going into the army as white. I think it had something to do with the idea that if we're sending people from our home county off to the army, we're going to send them in as white men so they'll be treated better, and we'll back that up at home. We're not going to have their kids go through a different school."

One might hope that the surge of Melungeon pride and the reclamation of a complicated, nonwhite identity would constitute a satisfying turn-about. But the situation is more complicated than that. Some Melungeons have themselves come to restrict the term so narrowly that it excludes most potential members. As one man said to Wayne Winkler, "If you can't trace your family back to Hancock County, you ain't a Melungeon. Period."

There is suspicion too about why people might wish to reclaim Melungeon heritage. Local people who have always identified as Melungeon are skeptical about "wannabes" who only now want to acknowledge a Melungeon heritage because it has become exotic or popular. While it's admittedly easier now for someone to call himself a Melungeon without having to suffer any of the explicit discrimination and shame that have historically burdened the group, the "wannabe" accusation is an easy but potentially crude label for people with unique motivations.

Winkler explained his own motivation for his interest in his ancestry:

> I want to document as best I can, the lives of those who struggled against racism and a rigidly enforced class system to survive. Those of us who descend from Melungeons owe much to our ancestors who worked hard to provide their children with a quality of life that they themselves would never enjoy.

For his part, Kennedy wrote about the effect of shame on many generations of his family. He had long wondered why so many of his people looked Mediterranean, why they often lived in inhospitable places, and why their surrounding communities treated them so badly. His own great-grandfather had not been allowed to vote, even in the twentieth century. No one in his family would explain any of this when Kennedy asked about it, or, if they did, their explanations struck him as unconvincing. When the topic came up, they often didn't look him in the eye. Only after tracking down piece after piece of evidence, many of which had been purposely hidden by his family, did he discover that he was Melungeon. The missing piece of his identity explained a lot of confusing incidents in his life, like the fact that as a girl his mother was always dressed in long sleeves, long skirts, and a hat, even in the summer—all to make certain she didn't turn "black."

The centuries of silence damaged his family, wrote Kennedy: "I saw the still-living tentacles spawned by this morass in much of my own behavior. This silent monster still lived and breathed and it had to be confronted if we were truly to move beyond it." He believed that reclaiming his heritage, coming out of the "Melungeon closet," would be a critical act of healing. His mother, who was uncomfortable with his choice, eventually grew to accept it. "I suppose it's like hearing a cry from the grave," she said, "and then having to decide whether or not to answer it."

Most investigations into Melungeon history are carried out by amateurs. Many are thorough, responsible, and compelling, but like much of the research that takes place in the sphere of genealogy and personal history, the lack of a university or corporate imprimatur leaves the area vulnerable to being dismissed as niche and unreliable. When Kennedy published his first newspaper article on the topic, long before the Internet and the lightning-speed responsiveness of services like Twitter, he received hundreds of calls and letters from people who felt they recognized themselves in his piece. Yet when he began his research, he wrote letters and telephoned many scholars to ask them for their thoughts, tried to fax his research to history and anthropology departments, and got no response.

Can genetics help lend validity to the stories of groups like the Melungeons? Currently the actual genetics are as complicated as the legends. In principle, if geneticists can identify the ancestry of large populations, they should be able to find ways to focus in on the more recent history of smaller populations. Indeed, the scientists who were able to detect differences in the small populations of Britain opened the door to this kind of fine-grained history. But as of now there have been only a few DNA analyses of Melungeon groups.

The largest study to date found evidence of male African American ancestry and female European ancestry, which is consistent with some of the legends. The subject group in this case was limited to descendants of those who had been described as Melungeon in historical records from the late nineteenth and early twentieth centuries. Yet records of the word "Melungeon" as a descriptive category for members of a particular population are rare and incomplete, although for a brief period at the end of the nineteenth century a number of individuals were noted as Melungeon in censuses. The subjects in the study represented just a small sample of families from Tennessee and other states. Moreover, within those families the researchers looked only at Y DNA and mtDNA, which in a set of possible great-great-grandparents represents just a small fraction of the DNA of two of the thirty-two people at this level.

What about the clustering of physical traits in Melungeon groups, the shovel-shaped incisors, the Anatolian bump, the palatal torus? Anthropology has long recognized that different physical features—such as the shape of one's head or teeth or the distance between one's eyes—occur at different frequencies in different populations. Richard Scott, a professor of anthropology at the University of Nevada at Reno, says that looking at teeth alone, he couldn't tell a German from an Italian, but he could always pick a German person from a Japanese or a Bantu person. Much like paper records and DNA, the evidence of the body can be definitive, but it can also be incomplete: The trick is knowing how to determine which it is.

The first step to determining if Melungeons developed a characteristic dentition would be to carry out studies to see if shovel-shaped incisors or other typical traits were significantly more common in Melungeon families than in the general population. If they were, it would suggest family connections, if not population-level ones. Here too we are only on the cusp of answers.

It's thought that one reason Melungeons might have shovel-shaped incisors is because Native Americans have shovel-shaped incisors. According to Scott, the trait occurs in 98 percent of the Native American population. If Melungeons have Native American ancestry, the trait may have been passed down from those ancestors. In fact, shovel-shaped incisors tell a story that dates back even further. More than fourteen thousand years ago, an extremely hardy group of people walked out of Siberia, across the Bering land bridge and down into North America. The first band that crossed what is now the Bering Strait originally came from Asia and brought their shovel-shaped incisors with them. The dimpled teeth are still a common trait in Asia and among Eskimo-Aleuts, and more than 90 percent of Chinese people have shovel-shaped incisors.

Shoveling does occur in Europeans and African populations, but much less frequently. Typically, it is the extent of shoveling, not just its presence, that distinguishes different populations. "For some reason, people always get fixated on shovel-shaped incisors," Scott said, "but they are only one of many traits that we look at." In fact, there are at least twenty-six different features of teeth that can help map ancestry. As far as shoveling is concerned, Europeans and Native Americans tend to be at opposite ends of the scale. With respect to other dental features, Europeans, Africans, and Native Americans are quite distinct. If Melungeons are a truly triracial population, there is a good chance it will be obvious from their dentition.

If we knew exactly how genes underlay these traits, and what particular genes they were, it would help piece together the history of the Melungeons. But the genetics of physical features is a nascent science. Simple traits that are shaped by one or a few genes are easier to identify. The moistness of one's earwax, for example, can be linked to a single letter within a single gene.

Many of our traits, however, are determined by several genes, with height being the classic polygenic trait. At least forty genes have been shown to contribute to it, and it's likely that hundreds of genes affect it. If that seems like genetic overkill, consider how many different parts of the body contribute to height: Someone may have a long shinbone or a long femur, or his spine may be longer than average, or he might have all of these features—it's likely that each of them is polygenic as well. Not even blue eyes and brown eyes are the straightforwardly Mendelian traits we used to believe they were. Despite what you probably learned in high school, it is possible for two blue-eyed parents to have a brown-eyed child.

Long before we sequenced the human genome or identified genes that contribute to teeth, anthropologists knew from tracking traits in families that many dental traits were polygenic. In 2011 scientists identified EDAR as the first gene known to contribute to shoveling. But we are still a long way from the full picture.

Of course, even when you are looking at genes, you are rarely looking *only* at genes. Some features are controlled by genes and *also* by whatever happens to be shaping them. We often think of genes as if they were master switches for the body—flick them one way, and you get blue eyes; flick them another, and you get brown. But genes can be influenced by a number of factors, including other genes, noncoding DNA, epigenetic markers (nongenetic attachments to cells), and chemical changes in the cell. The chemical changes are themselves often caused by larger systems in the body, which is, of course, shaped by the world in which it lives.

The palatine torus, for example, is shaped by development. "It has a pretty strong relationship to latitude," Scott explained, and is found predominately in Eskimo, Inuit, Siberian, and Native American populations. Scott also examined ancient Norse remains and found that it occurred frequently in medieval residents of Greenland. While it's clear that there is a genetic component to the trait, it's thought that mechanical stress, like chewing lots of cured reindeer meat, could cause a palatine torus to form.

Back in the nineteenth and twentieth centuries, Melungeons weren't ostracized because they had shovel-shaped incisors, as no one could see the backs of their teeth. They were treated differently because they *looked* different. The face is an incredibly important part of human culture and biology, and indeed, without becoming too circular, it is the major interface between the two. What happens on our faces shapes our initial encounters with others and our most intimate relationships. It is a powerful guide to internal states, and not just psychologically: A significant number of abnormal facial traits are correlated with defects in certain organs. Human brains even have specialized face-recognition mechanisms. Still, we are only beginning to understand how the human face is put together.

It's widely accepted that facial features are strongly determined by our lineage. Children often look like parents, siblings like each other, twins may look exactly alike, and even grandparents can look like their grandchildren. For the first two years of my eldest son's life, people in our neighborhood whom I didn't know would stop me on the street and exclaim, "Your baby is a clone of his father." It's true of our second child too. Baby photos of my son and husband, taken some thirty-five years apart, could easily be of the same child.

What makes family similarity so compelling and confusing is that it comes down to probability: Because you get 50 percent of your DNA from each of your parents, there's a good chance you'll resemble them both. There may be a reasonable amount of genealogical collapse in your family tree as well: If some of your great-great-great-grandparents who were married to each other were also first cousins, they brought more of the *same* DNA to the table when they had children. Their offspring are likely to have recycled more DNA from that lineage, and less variety was therefore passed down to you.

Then again, you *only* get 50 percent of your DNA from each of your parents. Sometimes people don't look like either of their parents or other members of their family. Some are tall where their parents are short or dark where their parents and siblings are fair. While appearance can be a persuasive pointer to ancestry, it is not always a reliable one.

Of the studies that have specifically looked for genes implicated in facial structure, most have attempted to determine the cause of an abnormality. In 2012 a group of scientists in the International Visible Trait Genetics Consortium published one of the first genomewide association studies to identify genes that contribute to the face. Led by Manfred Kayser from Erasmus University Medical Center in the Netherlands, the researchers took three-dimensional photographs of more than five thousand people and examined more than two million markers in the genomes of more than ten thousand Europeans.

The photographs were analyzed for forty-eight different facial characteristics, and the researchers tried to establish connections between the genomes and the different facial traits. They found five genes that contribute to the shape of the face. For example, the TP63 gene affects the distance between the eyes, while the PAX3 gene contributes to the distance between the eye and the nasion, the place where the nose meets the forehead. The findings suggest that, as with height, many genes contribute to the face's particular contours and size, each with a relatively small effect. Kayser is optimistic that much of the genetic basis of human facial variation will be discovered.

After the announcement of Kayser's groundbreaking findings, another team found evidence for the influence of many noncoding segments of DNA on the structure of the face. Soon after that, another team announced that their study using 3-D photographs of almost six hundred subjects had identified twenty genes that strongly influenced facial structure. Variations in one of these genes' codes were such useful predictors of facial shape that the team was able to build reasonable approximations of a subject's face based on his DNA alone. Overall, this kind of work may aid in the reconstruction of faces from ancient remains and ultimately even be used in forensic police profiling.

On the ancient meeting ground of Gulkula in Arnhem Land—the northeastern corner of Australia's Northern Territory—an elder called Gulumbulu demonstrated one way to span the worlds of whitefellas and blackfellas (Australian for white and black people): She was teaching her daughters and granddaughters the old ways while simultaneously sharing them with a large group of tourists—white Australians and visitors from other countries. Gulumbulu was a teacher at Garma, the country's largest alcohol-free, Aboriginal-driven festival of dance, film, and storytelling. Every day of the festival men and women streaked with white paint and wearing bright red wraps or yellow headbands danced on an open field on the edge of a large forest, which stretched many miles to Kakadu National Park. Huge mounds created by magnetic termites formed gray tombstones among the gum trees, and the woods were full of stinging green ants whose formic acid smelled like citrus. Beyond them an ocher escarpment dropped down to the inviting green Gulf of Carpentaria.

One morning I trudged up an unlit dirt road in the predawn hours with fifty paying tourists. We followed a group of elders to the edge of a cliff and sat quietly until the sun began to rise. This was a "cry for country" for women only, and the elders' gatekeepers warned, "No photos!" Just before dawn broke, birds began to sing, and the creaky-voiced Aboriginal ladies lifted their voices up as well, lamenting all the land that had been lost. It was sad and eerie and beautiful, until a spat broke out when one tourist furtively tried to take a photo and another crankily told her to stop. Yet another woman raised her voice and cried out with the elders. She was not one of them, and her skin was white. The women around her looked upset: Who was she to join in on the singing?

As we walked back from the morning ceremony, the woman who had sung told me about herself. She grew up in a poor family in New South Wales and was raised mostly by her single mother. When she left school, she sometimes traveled into small towns in the outback. More than once local Aboriginals singled her out from the group she was traveling with and said to her, "You are one of us." They welcomed her to spend time with them. As far as she knew, she was white, so she always took this to be not much more than a simple act of kindness. Much later she discovered that her mother was indeed half Aboriginal, a fact that had been concealed from her all her life. Years later, when she found herself on the cliff in the dawn light, she felt as if she belonged. The elders were the aunties she had never known, and she cried for her mother and her mother's father and everything that her family had lost as well.

We tend to associate some features with different kinds of ancestry. In the TV series *African American Lives*, host Henry Louis Gates Jr. discussed the DNA analyses of various guests, like the comedian Chris Rock and actor Don Cheadle. The show's guests described their family lore about Native American ancestry; relatives had made remarks like "That's Indian hair" and "You do have high cheekbones." Similarly, it's often said that there are only six faces in Ireland. The Y-chromosome data suggests that there is in fact a great deal of genetic overlap in Ireland, but we don't yet know how much all Irish genomes overlap, and if that might produce similar features. How much of the past is inscribed on our faces?

The Oxford geneticist Sir Walter Bodmer has long been interested in the face: "The fact that identical twins are so similar in their facial features, and they of course share essentially all their genetic make up, shows us that facial features must be largely genetically determined. The evolution of facial differences together with facial recognition must have been a very important part of the social and cultural evolution of the human species, and it's most probably connected with belonging and being recognized as a member of a group. The face also, almost certainly, plays an important role in choice of mate."

Bodmer and his colleagues are currently investigating faces in their study of British regional genetics. He explained, "It is a common observation that there seem to be facial characteristics that are associated with particular regions or countries even when they are basically closely related, such as within Europe. There are, of course, very obvious differences between major ethnic groups, such as Europeans and East Asians."

But this raises a question: If the genomes of the small British groups weren't distinctive enough to look different in a medical study, how could they produce different types of faces, even subtly different ones? "Even within Europe there has to have been sexual selection with respect to facial features, and that means, on the whole, picking people who are somewhat similar. I think that's been a very powerful force during evolution," Bodmer said. His team is in the process of taking three-dimensional photographs, each with 3,500 points of reference ("a full canvas"), of subjects who participated in the first stage of the study.

In retrospect the woman at the "cry for country" ceremony believes she should have realized she was not actually white, especially because of the way other Aboriginals treated her. Yet no one else in her life had guessed her ancestry. She looks like her mother, who looked like her own father, who was recognizably Aboriginal. Could other Aboriginals tell that she was part Aboriginal because they were Aboriginal themselves?

It turns out that some people are better than others at making judgments about ancestry based on looks. For decades psychologists and anthropologists have investigated a phenomenon called own-race bias. At least forty different experiments have demonstrated that people are better at remembering faces when the face appears to be the same race as their own. This is true no matter what the race of the observer or the observed is. It's also been shown that people are better at predicting how well they will perform a face-recognition task when the race of the photographed face is thought to be the same as their own, which is to say we overestimate our ability to judge how well we recognize faces from races other than our own. It's not entirely clear what the mechanism responsible for own-race bias is. One of the most important implications of own-race bias is that in eyewitness situations the testimony of a witness may be considered less reliable if the accused is of a different race.

Most studies of own-race bias have relied on self-reported race. In 2012 Mark Shriver, an anthropologist at Penn State University who studies the interplay of faces and genes, ran an experiment that investigated the connection between genetic markers of ancestry and the ancestral cues we detect on people's faces. He asked more than two hundred subjects who lived in New Mexico to assess the component ancestries of fourteen Hispanic faces based on front- and side-view photographs. The genomes of the people in the photographs had been analyzed with respect to their mixture of Native American, European, African, and East Asian ancestry.

Shriver found that most observers made a better guess at the admixture of the photographed individuals than someone who simply guessed randomly. Still, the guesses were far from perfect, suggesting that, while we have some general ability to detect ancestry, it's not uniformly reliable. Shriver's results were consistent with other studies that show the more similar the observer's ancestry is to the ancestry of the person in the photograph, the better the observer is at guessing the correct family history. The most plausible explanation for this is that the observers had *learned* to interpret the features of faces with which they were most familiar.

Human skin color is another important case study in the way that genes shape traits. An inherited trait, the many varieties of skin color have emerged in a surprising way from the intersection between environment and behavior in the last hundred thousand years. When humans left Africa and began to live in the Northern Hemisphere, the color of their skin became lighter. This has long been attributed to natural selection and the need for skin to make vitamin D. As the details of human genetic history are being uncovered, some changes look more like a case of *relaxed* selection rather than natural selection. Many genes are involved in skin pigmentation; the MC1R gene (melanocortin 1 receptor) is critical in the production of melanin, which darkens the skin. In Africa today at least eleven different mutations to MC1R have been identified, but eight of these are so-called synonymous mutations, which do not actually affect the related amino acid in the protein structure or the protein's function. The fact that most of the African MC1R mutations are synonymous means that what the gene does is critical in that particular context. Outside of Africa MC1R has undergone many more mutations, and many of them do have an effect on melanin production. So what has been hard won by positive selection in the bright light of Africa—strict genetic control of dark skin that protects against damage from strong ultraviolet radiation—disappears when it no longer affects survival. Outside Africa it appears there are many ways to turn white. The nonsynonymous mutations to MC1R are diverse and vary from region to region, but most lead to the same result: reduced melanin production. Some mutations merely alter the function of MC1R, while others completely shut it down. The red hair and freckles of many people in the British Isles results from just such a mutation.

Skin pigmentation is affected by more than genes. It may even be that lighter skin was inherited from Neanderthals, as regions of the genome that contribute to skin color show influence from the Neanderthal genome. Yet change did not happen only in the distant past. There is evidence from ancient DNA that lighter skin, hair, and eye pigmentation was strongly selected for in Europe in just the last five thousand years. The change could have resulted from the greater success of people who were able to process more vitamin D, or it could have resulted from sexual selection, where people with lighter pigmentation were more successful in reproduction.

There is much to be learned about how DNA shapes traits and how traits then shape our experience of the world, either because of our abilities or because of the way people treat us. One of the most crucial aspects of this nexus is the way that genes affect our well-being, either by predisposing us to disease or by protecting us from it. As with everything else to do with DNA, the forces of fate and randomness play a huge role in people's health, and as always the family is often the crucible of this drama.

Chapter 14

The Past May Not Make You Feel Better: DNA, History, and Health

> The laws of genetics apply even if you refuse to learn them.
>
> —Alison Plowden

When Jeff Carroll was sixteen he dropped out of high school. At twenty he joined the army and was posted to Europe. He served in Germany for a year, and on his first trip home for Christmas, his father told him that his mother was showing signs of Huntington's disease, a condition that Jeff had never heard of. Huntington's is the cruelest diagnosis. Patients slowly lose control of their bodies, as well as their memories and their ability to think. They may undergo personality changes too, often becoming aggressive toward their loved ones. The degeneration is slow and relentless, unfolding over the course of years. Although Cindy Carroll was in her midforties when her body started to jerk without warning and she forgot one of her best friends' names, she lived for many years afterward.

After his father told him about the diagnosis, Jeff went back to his life in the service and later enrolled in an army biology course. When he returned to civilian life, he completed an undergraduate degree in biology and began working toward a PhD in a lab studying Huntington's. During this period he got married, and in the last year of his mother's life he and his wife had twins.

By that stage Cindy Carroll's body was so constantly overcome with the uncontrolled jerking and writhing, called chorea, that is typical of the disease that all her nursing home could do was place her on floor mats and hope she wouldn't hurt herself. The night she died, Jeff brought his baby son to her, carefully placing him in the crook of his mother's neck. It had been years since Cindy had shown any sign that she recognized her son, but when the baby nestled in to her, she was briefly still and seemed at peace. Her respite probably lasted only a minute, Jeff said, but to him it felt like hours.

After his mother's death Carroll told a reporter that the worst thing about the disease was not the fact that it is fatal but that it "destroys your personality and turns you into an object of horror for your family." Yet that is not the end of it. Huntington's disease is hereditary, and when people talk about things like destiny and genetics and whether it is wise or not to know how you will die, Huntington's is often what they have in mind. Cindy Carroll died in 2006, six years after her own mother, who was also a Huntington's sufferer. When Cindy was first diagnosed, Jeff and his siblings learned that because of the way the Huntington's mutation works, they had a 50 percent chance of developing the disease themselves.

In 1993 researchers identified the genetic mutation that causes the disease. The discovery made an enormous difference to the likelihood that a cure would be found, and it led to the development of a test that can determine if someone will develop the disease. Before the test existed, the children of Huntington's patients could only watch the suffering of their family members and wait anxiously to see if they developed symptoms, asking themselves whenever they dropped something if it was clumsiness or Huntington's. Now the test brings grim certainty: If candidates have the mutation, they will develop the disease. Yet more than 80 percent of the people who could take the test do not.

Jeff always knew he wanted to be tested, but it wasn't until 2003 that he started the process. On July 31, 2003, he and his wife met with a physician to get the results. The physician told Carroll that he was positive for the mutation.

Descriptions of a disease that sounds like Huntington's can be found in writings that date from the Middle Ages. The uncontrolled twitches and swooping, circular, constant motion of the Huntington's sufferer were first described as a dance in the 1500s. Later observers drew closer to understanding the condition and in the nineteenth century finally connected the affliction in one person to a similar condition in one of his parents. In

1872 the young physician George Huntington was the first person to clearly describe the illness as hereditary and degenerative, with an onset typically taking place in the afflicted in their thirties. "Those who pass their fortieth year without symptoms," he wrote, "are seldom attacked."

Almost two hundred years later the science of Huntington's was forever changed by Nancy Wexler, a thirty-three-year-old New York–based neuropsychologist whose mother was diagnosed with the disease in 1968. "It was as if some mad puppeteer was in control of her body," Wexler wrote. In 1979 Wexler traveled to a small town on Lake Maracaibo in Venezuela to visit the largest Huntington's family in the world. Since the 1950s the people who dwelled in the town by the vast and ancient lake had been known in the medical literature for their extraordinary one-in-ten chance of developing Huntington's, which they called *el mal* or "the bad."

Wexler founded the U.S.–Venezuela Collaborative Research Project and for several decades traveled to the region every year, studying pedigrees and sampling blood. She worked out that the villagers of Lake Maracaibo had seen more than eighteen thousand cases of Huntington's in the span of ten generations. There are a few stories about when, and with whom, Huntington's began in Lake Maracaibo. Some say the first cases were children of a woman called Maria Concepcion who lived in the early 1800s. Concepcion had ten children, and it's thought that their father may have passed on the mutation. Genealogical work has traced tens of thousands of cases of Huntington's to Concepcion's pedigree. Another story, probably apocryphal, is attributed to a physician who diagnosed the villagers' condition as Huntington's in the 1950s and wrote that the locals told him that sometime between 1862 and 1877 a ship's priest named Antonio Justo Doria left his ship and decided to live by the lake. He married and had children, and later in his life he was seen "walking with some strange movements, like dancing." Currently one thousand locals have the disease, and around five thousand are known to carry the mutation for it. On a 2010 visit Wexler encountered a single large family in which both parents and ten of their fourteen children had Huntington's.

In 1983 Wexler's team got close to the gene when they found a marker that was closely linked to it. In 1993 they finally identified the "hunting-tin" gene, as well as the mutation that caused the disease. Because the Huntington's mutation is dominant, you only need one mutated copy from either parent to develop the disease. As categorical as the disease is— you either have it, or you do not—the genetic underpinnings of Hunting-ton's are oddly not so exact. Huntingtin contains a repeated sequence of the letters CAG. In normal copies of the gene, CAG is repeated around seventeen times, and it can be repeated up to twenty-six times with no obvious consequence. However, if the CAG sequence is repeated more than forty times, the carrier of that gene will develop the disease. When Carroll was tested, he found out that he had forty-two CAG repeats.

While forty repeats is a definitive threshold, the CAG repeats have an odd additive effect as well, which people in the community call the "gray area": If you have between thirty-five and thirty-nine CAG repeats, you will get the disease, but it won't strike until your seventies or later.

If you have between twenty-six and thirty-four repeats, you will not develop Huntington's yourself, but there's a small chance that, if the gene you pass on mutates further, you may have a child who does. Even though Huntington's is so strongly hereditary, Carroll explained that 10 percent of the new cases every year occur in families where there is no history of it. Initially people suggested that such instances might be cases of adop-tion or illegitimacy, but that was shown not to be true.

Huntington's usually appears in its sufferers between thirty and fifty years of age, but in rare cases children may also display symptoms of the disease. Often people with Huntington's develop symptoms around the same time their parent did. But there's a tendency too for it to appear a bit earlier if the mutation was inherited from the father. Wexler showed that the more repeats someone has, the earlier he will get the disease. The highest recorded number of CAG repeats on a huntingtin gene was near one hundred, a mutation carried by a boy whose symptoms began when he was two years old.

Huntington's may be the starkest model we have for reflecting on biology and fate. It's a Mendelian disease, which means that the condition arises from a single gene.

Huntington's also has deep resonance for how we think about all DNA. While so much knowledge has been steadily acquired in the realm of genetics over the last century and a half, and so much brilliance has blazed in the field in the last twenty years, there is still more dark matter in this particular universe than not. "We just learned the alphabet," observed Carroll, "and we were claiming we could write Shakespeare, and it's a long way from here to there."

The scientific and citizen community that has formed around Huntington's disease is by now a highly educated one. The discovery of the gene had enormous consequences not only for Huntington's disease but for all genetics. The genetic test for the mutation was the first offered for a genetic disease that appears in adulthood. Some of the techniques developed to identify huntingtin were later utilized to sequence the human genome. Still, despite their intimacy with even the smallest molecules that affect the gene, there is a long list of incredibly basic questions to which scientists do not yet have the answer.

Scientists, for example, are baffled by what happens to someone who has two copies of the mutated huntingtin gene. It's an extremely rare occurrence, but sometimes a man with Huntington's and a woman with Huntington's will have a child. That child will have a 75 percent chance of inheriting one mutated copy of the gene and a 25 percent chance of inheriting both. Yet despite the fact that more repeats on a single mutated copy means Huntington's symptoms have an earlier onset, somehow people with two mutated copies of the gene do not develop a more severe set of symptoms than people with just one copy.

When Carroll gives a presentation about Huntington's, he sometimes shows his audience a picture of slime mold, because slime mold has a huntingtin gene too. If a creature as simple as mold has a gene that humans also carry, then we can assume that a shared ancestor, an entity that lived millions and millions of years ago, had it as well—which means that all the creatures on the great evolutionary tree between mold and humans likely have it too. Any gene that is conserved in the genomes of many creatures is maintained because it has a very basic and very important function. The Hox genes, for example, are shared by all vertebrates and control their basic body plan, a central spine from which limbs project on both sides (in comparison to, for example, the blobby, spineless jellyfish). Yet scientists do not know what the function of huntingtin is in humans.

The significance of the huntingtin gene is demonstrated quite vividly by slime mold. When researchers turned off the gene in mold, it became sick. But remarkably, the huntingtin gene in mold and the huntingtin gene in humans are so similar that, when researchers put a healthy human huntingtin gene into the sick slime mold, it got better.

Not only is the huntingtin gene extraordinary for its spread across species, but its reach within the body is amazing too. Typically genes produce proteins in particular cells but not in others. It's also generally true that genes produce proteins for a particular period of time but then stop doing so: They are turned on and off in the normal course of development. Huntingtin, by contrast, is one of the rare genes that is expressed in all tissues all the time. The protein the gene expresses is also called huntingtin, and it can be found in the cells of the heart and the lungs, in the blood and in the brain, and in the bones. Yet scientists still don't know what the protein actually does. "It's not super dynamic," explained Carroll. "It doesn't seem to change its expression levels in response to signals, which a lot of other genes do. It's like a housekeeping thing, it's always there."

Carroll is a tall, clean-cut, strawberry blond who looks as if he still trains with the army. When he is introduced at conferences, presenters joke about how handsome he is. (In 2012 one colleague welcomed him to the stage by saying that during the Kosovo war, the women on both sides persuaded their husbands to lay down arms, just so they could look at Carroll.) He got his big break after his undergraduate degree with a job in the laboratory of prominent clinician and researcher Michael Hayden in Vancouver. Hayden's team was trying to develop a drug that could silence the huntingtin mutation. "Gene silencing in Huntington's is really attractive," Carroll explained, "because of the fact that it is a Mendelian disorder, so 99.9 percent of people who have Huntington's disease have the same mutation—variable length but the same place—and vastly all of them have one good copy and one bad copy of the gene." The drug in question would essentially be a "short piece of DNA or RNA" that shuts down the bad copy.

Most research on gene silencing and Huntington's has been focused on pan-huntingtin silencing, meaning that both the mutated copy and the nonmutated copy of the gene would be shut down. It's easier to begin with this goal, Carroll explained, but it can take researchers only so far. Mice that have both versions of the huntingtin gene silenced in utero are nonviable; mice who have the gene silenced when they are older do better. But it's still unclear how humans would fare with such a treatment.

The best way to silence the gene would be to target only the mutated copy, which was the focus of Carroll's work. His team found that some of the letters of DNA in noncoding regions near the mutated gene were closely correlated with the mutation. By using those letters as a kind of address for the mutation, they were able to silence the bad copy in laboratory tests. In other tests that knocked out the mutant huntingtin in mice, one unexpected consequence was a rebound effect: The mice not only stopped deteriorating but actually got better. "Scientists are calling it a 'Huntington holiday,'" said Carroll. These drugs may not be able to stop the progression of Huntington's forever, but they may give the brain "some space to compensate for some of the damage that it has experienced."

Carroll put in long, hard hours on the drug, but once it reached the point of testing, he found he wasn't motivated by the careful, plodding work of safety trials. "I got off my high horse," he explained. "You get humbled by therapeutic development, you realize you're not that important. This is a team effort and I didn't have to be the guy at the pointy end of the pipeline. I could decide what I wanted to do with my life."

Carroll is now working on Huntington's and metabolism because he became fascinated by "the remaining mysteries." For example, even though the huntingtin gene is expressed everywhere, the places in the body where it is expressed more aren't the ones that are most damaged in the course of disease. While most research examines the effect of huntingtin on the brain, because it so obviously and dramatically degenerates, Carroll is fascinated by changes in the liver and pancreas and other tissues caused by the mutant gene. "If you have cirrhosis of your liver, you get profound neurological symptoms." In some respects, neural imaging in these cases looks a lot like those of Huntington's disease. "So you don't have to just have brain degeneration that causes a brain disease; you can have peripheral dysfunction that causes brain dysfunction."

He is also exploring the problem of Huntington's and food. "It's well known among caretakers and caregivers that Huntington patients eat a ton," Carroll said. Some Huntington's sufferers consume as much as five thousand calories a day just to maintain weight. "They're hyperphagic, and yet they lose weight," he said. Many of them die of starvation, but, as Carroll explained, "Nobody knows why."

Carroll also started a Web site called HDBuzz with a colleague, Huntington's clinician Ed Wild. Both men were concerned about the amount of misinformation and hype about Huntington's in the press, and they were struck too by the fact that while affected families desperately needed up-to-date information about research on the disease, Huntington's also desperately needed affected families to help them with their studies. The site helps the two connect.

When Carroll discovered he carried the huntingtin mutation, he decided he would never have children and risk passing the mutation on, but he changed his mind in the early 2000s when scientists came up with a method to ensure that the child of someone with a mutated huntingtin gene would never inherit that copy. In fact, two methods were developed. Doctors can test a fetus early in pregnancy and terminate it if it carries the mutation. They can also conduct a preimplantation genetic diagnosis. Using either a couple's egg and sperm or a donor's, doctors can create and test embryos for the mutated gene and then implant a noncarrier using in vitro fertilization. Jeff and his wife, Megan, were in the first generation of couples to use the second method, and after one try they conceived nonidentical twins who do not carry the huntingtin mutation.

It used to be the case that most people started families before they knew they had the disease. Often they had children before they themselves began to display symptoms, and sometimes even before their own parents developed symptoms. There was also a great deal of secrecy about having the disease, a tendency to hide a diagnosis or not discuss it. In some cases the symptoms were thought to be caused by something else, like alcoholism.

Despite the availability of testing, at least half of the population at risk for Huntington's disease still has children without making use of the new technologies. Even some of the people who have prenatal testing for Huntington's still have a profound reluctance to learn their own status. Couples who try preimplantation genetic diagnosis may even conceive a child and choose not to find out if the parent at risk has the mutation.

Deciding whether to find out one's genetic status is a hugely divisive and painful issue within families and the Huntington's community, but the desire to conceal information from oneself is a fraught position, especially when other people are affected by the same genes and the same knowledge. In one Huntington's family a young woman was discouraged from taking the test by her mother, who did not want to know if she herself carried the mutation. The daughter eventually distanced herself from her mother, took the test, and discovered she was positive.

Silencing the mutation will put an end not just to the disease but to all the associated issues around disclosure. Until that happens, though, the issue remains extraordinarily stressful. Many adults who are at risk for Huntington's and who have not taken the test worry that if they receive a positive result, employers may learn of their condition or, worse, insurers will. Although various organizations lobby to prevent genetic discrimination, it's unclear how policy will progress, as the science changes so rapidly (more on insurance issues in the epilogue). Mostly, though, it seems to be the case that at-risk adults are traumatized and weary from looking after afflicted family members, and they feel terrible anxiety at the prospect of developing the disease themselves. The time just before people receive a formal diagnosis of Huntington's is a critical period for suicide. Even those who are negative for the mutation may be haunted by survivor's guilt. For most it is better to imagine that they do not have the mutation than to seek the knowledge that they don't and risk finding out that they do.

Carroll is part of a much smaller group who have taken the test. "Some people just have to know," he explained. And he is also part of an even smaller group: scientists who are at risk for Huntington's who devote their careers to understanding the mutation. He thinks he can make it to forty-nine years of age before serious symptoms start to appear. In the meantime he has work to do.

Huntington's may be the prototypical example of a Mendelian disease, but not all single-gene disorders are the same. Indeed, there's a long list of ways that a single gene can have an impact on health and that a genetic test can change a life.

The Samaritans are members of an ancient religious sect who live in the village Kiryat Luza on Mount Gerizim in the West Bank and in a town called Holon in Israel. They have the highest rate of inbreeding in the world.

In Roman times there were one and half a million Samaritans. According to their history, they descend from the sons of Joseph and lived in the northern kingdom of Israel in the Solomonic period, around 1000 BC. In the early eighth century BC the Assyrians swept through, exiling or supplanting many of the inhabitants, yet somehow the Samaritans were able to remain. When the Israelites returned from exile, they rejected the Samaritans from the tribes of Israel because the Samaritans had adopted some Assyrian customs. Even today, according to geneticist Marcus Feldman, who studied the group, Samaritans are not considered Jews. Yet work carried out by Feldman's lab has shown that their ancestry is very similar. The genomes of the Samaritans, he said, "seem to be very close to those of other Jewish people after all this time."

In the centuries after the Assyrian invasion, one catastrophe after another befell the Samaritans, and in the wake of attacks by Romans, Muslims, and Ottomans, the population progressively shrank. By 1917 there were less than 150 Samaritans left. In the years since then the group has slowly recovered from its extreme bottleneck. As of 2009 there were approximately 750 Samaritans.

One of the reasons the population has struggled to expand is its higher-than-average risk of certain genetic disorders. For much of the twentieth century Samaritans had relatively high rates of miscarriage, stillbirth, serious disability (such as being deaf or mentally retarded or unable to walk), and infant mortality from degenerative diseases. These health issues are in part a consequence of the community's commitment to wed only other Samaritans. Marrying a close relative increases the risk of a genetic disorder, because if there are recessive mutations in a larger family group, when both parents come from that group the likelihood that they both carry the recessive mutation is increased. But the Samaritans are unique in their preference to marry not just within their larger group but within their own surname group. There are four surnames in the entire community, which shrinks the genetic pool considerably.

At least 84 percent of Samaritan marriages take place between first and second cousins. A marriage between first cousins might increase the risk of some conditions, Feldman said. But it is the repeated choice to marry first cousins over many generations that vastly increases the risk of a genetic disorder. Simple math tells us that ten generations ago in anyone's family tree, 1,024 people coupled up. All 1,024 people are genealogical ancestors, but if they contributed anything at all to the genome of their modern descendants, it was only a small amount of DNA. Reality, though, is often more complicated. The math is correct *only* as long as no couple in the family tree was related to his or her spouse. When there is repeated marriage between cousins in a small population, the amount of DNA shared by any two spouses increases and the number of genealogical relatives in the tenth generation (as in all others) decreases. As the number of genealogical ancestors shrinks, the chance of inheriting a DNA segment from any one of them increases. While it seems improbable and almost infinitesimally unlikely that any one child would inherit two copies of a mutated gene that has been passed down over multiple generations, when a community prefers endogamous marriage, the odds drop.

In a marriage of first cousins, for example, the children will have six great-grandparents instead of eight. The children of this first-cousin marriage will have fewer than 1,024 ancestors in the tenth generation. In such a marriage it is necessarily the case that two of the married couple's four parents are siblings. In a double first-cousin marriage, the children will have four great-grandparents instead of eight. If you repeated the pattern many times over throughout the generations, the size of the ancestral pool would shrink considerably.

It is extremely unlikely that anyone in the twenty-first century does not have some consanguinity in his or her family within the last three hundred years. Yet according to Feldman, more than half of all human populations today still engage in consanguineous marriage, and up to 10 percent of all humans are in first- or second-cousin marriages.

The Ashkenazi population is much larger than that of the Samaritans, but it went through a bottleneck between the tenth and fifteenth centuries after the Ashkenazis had been expelled from France and the Rhineland. Even though it has expanded into a community of ten million people worldwide, it's thought that all Ashkenazis are at least ninth cousins to one another. The physical legacies they deal with today include Tay-Sachs disease, a recessive, degenerative disorder that is often fatal by four years of age. In the United States typically one in 250 adults carries a recessive copy of the gene. But in Ashkenazi communities that drops to one in every 27 people. There are at least twenty genetic diseases that Ashkenazis are more likely to be afflicted with than many other populations. Their suite of risks, said Feldman, is shaped by the fact that they are a small population with a preference for marrying people within their own communities, and the founder effect.

The founding fathers and mothers of a population may have enormous influence on their descendants. Consider: The founders of a small group are just a random sampling of their original population. The sample may be a small-scale representation of the diversity in the original population or, more likely, it may be a tiny subset of the genomes in the population they left behind. If one of the founding group has a recessive mutation, and it's customary to marry within the group, it's likely that with a few generations, distant cousins who are both carrying a copy of the same mutation will marry. Tay-Sachs disease results from a mutation on the HEXA gene, and it's thought that the founder who introduced it into the Ashkenazi population lived during the fifteenth-century bottleneck.

Disorders like Tay-Sachs, Gaucher's disease, and Bloom syndrome are genetic risks for the Ashkenazis, but they are not exclusive to them. Other populations are also at higher risk for some of these disorders—or for conditions that the Ashkenazis themselves tend not to get. The Irish, French Canadians, and Cajuns also have a higher incidence of Tay-Sachs disease than do other groups. None of these groups practiced cousin marriage, but they were small populations in which the locals, as a matter of course, preferred to marry people who were like them or who lived close to them. Curiously, a genetic disorder shared by two or more populations doesn't necessarily imply shared ancestry for those populations. While it is often the case that a particular mutation on a gene may have a dire consequence and a different mutation to the same gene have no apparent consequence at all, sometimes the same disease results from different mutations to the same gene. French Canadians, for example, carry a different HEXA mutation than the Ashkenazis, one that has been traced to a carrier in the seventeenth century.

Cajuns, however, do carry the same mutation as most of the Tay-Sachs-affected people in the Ashkenazi population. Until the nineteenth century Cajuns were relatively isolated, and there was a high degree of marriage within extended families. Even in the twentieth century many families stayed in the same areas for generations, and some of the signs that individuals there might be related, like common surnames, were lost in time.

In the late nineties Iota, a small town in Louisiana, experienced a disturbing spike in Tay-Sachs diagnoses. Within a period of months four cases of Tay-Sachs disease were brought to the attention of New Orleans clinician and geneticist Emmanuel Shapira. When more cases followed, Shapira began to try to trace the recessive gene. He visited Iota and, after taking 230 blood samples, found that the percentage of carriers of the Tay-Sachs mutation was twice that of the Jewish population. Where had it come from? Shapira and his colleagues examined the pedigree of seven Cajun families affected by Tay-Sachs, all of whom lived within seventy miles of one another. Most of the families were found to share common ancestors: a couple who immigrated to Louisiana from France in the early 1700s. Of the seven families in the study, five were traced directly to the couple; the other two were traced to individuals with the same name who lived around the same time. It's likely that they too were related, but a definitive connection to the original couple could not be established. It's not known whether the eighteenth-century couple was Jewish or not, but some commenters have speculated that they must have been.

The common ancestry in the families suggests that multiple copies of the same segment of DNA, which happened to contain the mutated HEXA gene, came from one person who lived around three hundred years ago. The mutation was copied and recopied within the Cajun community, whose members by the late twentieth century no longer knew if or how they were related to one another. Indeed, they weren't related in the way we normally think of it. They possibly didn't have much other DNA in common at all, only the fateful HEXA segment. But of all the parents in the entire Cajun family tree, that one couple played a uniquely important role in their lives.

Why after hundreds of years of isolation did so many young adults each carrying a single copy of the recessive gene unwittingly pair up? In fact, it's likely that previous generations also did so, and indeed, once the extended families of the afflicted children learned about their condition, a number of them recalled similar cases in previous generations where an otherwise normal infant stopped thriving, began to degenerate, and eventually died by the age of four. Locals used to call the condition "lazy baby disease."

A community's experience of its genome will be shaped not only by its genome, but by the technology it has access to. Unsurprisingly, the Samaritans' experience of their own genome and genetic information is significantly different from that of people in the Huntington's community. Once individuals with the huntingtin mutation, a disorder of just one copy of a gene, learn their diagnosis, there is nothing they can do to change it. In the case of the Samaritans, who typically face recessive disorders, which require two copies of a gene to be mutated, steps can be taken. While there are no cures for the genetic disorders that afflict them, the adult carriers of recessive genes are not affected, so their focus is on prevention in the next generation. Samaritans actively seek information about the genomes of potential offspring. They now take part in both premarital and prenatal testing, and even though one in every five pregnancies is abnormal, they can determine which embryos carry two copies of the same mutation and choose to terminate. In this way they change neither their genome nor their cultural practices, yet testing ensures that only nonaffected children are born.

As well as genetic testing, a small group of Samaritans have used specialist agencies to find wives from outside the sect. In the last decade a number of young Ukrainian women have been recruited to marry Samaritans, bringing with them entirely new genomes in order to maintain the three-thousand-year-old culture. Still, many Samaritans remain committed to marrying within their families. "I'm against marrying women outside our community," one Samaritan told a Reuters reporter in 2009. Speaking of his sons, he said, "If they don't find a wife, my sister has three daughters, and my cousin has three daughters." He added, "Of course, we'd have them tested genetically first."

Such serious engagement with genetic testing has become prevalent in the larger Jewish population over the last twenty years. In Israel genetic screening and counseling are a normal part of the culture. Everyone is screened for fragile X syndrome, and other tests are offered free to at-risk couples. Mutations leading to rare but devastating genetic disorders are still being discovered as well. In 2012 the mutation causing progressive cerebro-cerebellar atrophy, a fatal degenerative childhood disorder, was identified and the Israeli government moved to test for it as part of a larger battery of tests.

I asked Marcus Feldman, who is himself Ashkenazi, if he was worried about having children. "Not at all," he said. "Once you get past second cousins, the danger of producing genetic diseases drops to numbers that are very small."

The risk is "significant enough that one would probably want to do some prenatal testing," he explained. "But I don't think people do a routine Blooms disease test because although these diseases pop up in the Ashkenazi community, they are very, very rare even among Ashkenazi."

In the United States organizations like Dor Yeshorim in the ultraorthodox Jewish community carry out premarital testing. If results reveal

that both individuals in a couple are carriers for the same recessive condition, then approval is not given for marriage. Dor Yeshorim and programs like it are so successful that there are now fewer cases of Tay-Sachs
disease in their communities than in non-Jewish ones. In the United
States and Canada cases of Tay-Sachs have been reduced in the Jewish
community by more than 90 percent since 2000. Here is a clear case
where genetic risk factors have become decoupled from cultural risk factors, and the culture has adapted so as to diminish the genetic risk. Currently public-health information and genetic testing in the Cajun
community are less developed. Many twenty-first-century descendants
of the original French Louisiana couple and other local families with a
different history may still carry the Tay-Sachs mutation but not be aware
of it, and they may yet have children who are afflicted. Or, if they happen to marry someone who is not a carrier, they will not have to know
their own status or deal with Tay-Sachs.

In other communities that carry the legacy of founding genomes or
cousin marriage, targeted screening programs exist. Many countries, including Canada, Cyprus, and Iran, have screening programs for beta
thalassemia, a blood disorder that severely impacts development and
may require a carrier to have lifelong blood transfusions. The countries
differ in whether testing is voluntary or mandatory, prenatal or ante-
natal, and in what kind of counseling is offered. Couples in Cyprus must
be tested and issued a certificate if they wish to marry in the Cypriot
Orthodox Church. In Cyprus, and possibly Canada and Bahrain, the incidence of beta thalassemia has dropped near 90 percent. In other countries, such as India, little or no progress has been made.

The issue of marriage and genetic testing can be extremely culturally
sensitive. There was much controversy in recent years when it was announced that a Pakistani community of two million people in Bradford,
England, had a one-hundred-times greater frequency of genetic disease
than the general population. The community had married within its clan
for many generations before immigration and continues to prefer marriage between first cousins. Now one in ten of its children develops a
recessive disorder or dies in infancy. In an interview on a British television program a local doctor estimated that while other hospitals would

normally see 20 to 30 cases of a recessive disorder a year, the Bradford hospital sees around 140 cases. The British government has declined to address the public-health issues in Bradford in any systematic way, and there is much anxiety about criticizing cultural practices that are relatively new to the country. Some insist that first-cousin marriage is not a government issue; other medical and political figures are debating ways to address it.

Alan Bittles, one of the world's experts on consanguinity and author of *Consanguinity in Context,* became interested in the subject when he visited Bangalore to do research in the 1970s. At dinner with his professor, the man introduced his family to Bittles. He said, "This is my wife and she is my niece." Bittles also met the couple's children. "They were bright attractive kids," Bittles said, and it made him wonder about the dire warnings that the medical community of the time attached to consanguinity. The vast majority of first cousin marriages do not have children with birth defects, he explained. Fundamentally, what matters is not just the consanguinity but the size of the group, how many children they have, the degree to which the group's ancestors were isolated, and in many cases, socioeconomic factors like maternal education and maternal age. There are many varieties of consanguinity as well, each with their own impact on the genome of children. Uncle-niece marriage, which is very common in Bangalore, for instance, is twice as inbred as first cousin marriage. For this reason, consanguinity is best considered as a "spectrum"—it depends on how many identical segments of DNA both parents have inherited from a common ancestor.

Any population, no matter how large or small, may have a greater predisposition to some disorders than to others. While the populations of western Europe and West Africa can hardly be considered isolated, they are still home to large ancestral groupings of genomes that may affect their carriers' lives. One in two thousand western-European births is affected by cystic fibrosis, whereas the condition is rare in Africans. West Africans, on the other hand, must deal with sickle-cell anemia in one of six hundred births, but the condition rarely appears in European populations.

The amount of shared DNA within a population isn't affected just by cultural choice or bottlenecks from medieval or colonial times. Even to-day we are shaped by an event that began sixty thousand years ago: the out-of-Africa journey. When humans left Africa and migrated all over the world, they passed through a series of bottlenecks as one population settled and expanded and then a small group broke off and moved on and founded another population. Brenna Henn, a colleague of Marcus Feldman, led a study that found that for every population bottleneck there was a decrease in genetic variation and an increase in deleterious mutations in the genome. Currently Feldman and Henn are investigating whether these impact the health of individuals today.

Millions of people are affected by Mendelian diseases. One birth in every thousand is shaped by a single-gene disorder, of which there are at least ten thousand. Yet as enormous as this number is, Mendelian diseases are considered rare. We have known since Francis Galton, the great innovator, eugenicist, and cousin of Charles Darwin, that not everything we inherit can be explained by a single gene. Now, as the revolution in genomewide association studies enables us to compare the genomes of many people, we know it to be true that *most* genetic effects are caused by more than one gene. Many traits and common diseases cluster in families, so we should be able to find traces of them in the genome, but so far those traces have been elusive. Some rather large factor is missing from the picture.

The great paradox of this last crank of the scientific wheel is that while we have finally affirmed the principle that many genes may contribute to one condition, we still don't know how all those genes do what they do. Now that we have the technology to establish how much influence any one gene has, it *appears* to be the case that a lot of the time they don't have that much influence at all.

Take height as an example. In families height appears to be a strongly inherited trait. In addition, genomewide association studies indicate that it is influenced by at least forty different genes. Yet when scientists tried to understand how these genes underlie the pattern of inheritance, they couldn't work it out. The height genes were found to only explain 5 percent of the difference in height in the population. Clearly, a lot of *something* happens between the genome and the person, but as of now we don't understand precisely what that is. Geneticists call this the problem of "missing heritability."

The first explanation, naturally, is other genes. If a condition is the product of many genes, it seems likely that the activity of some genes or pieces of noncoding DNA may affect the actions of others. It may also be the case that some common diseases are shaped by rare mutations that have not yet been tracked down. As advances in medical care keep many more of us alive than hundreds of years ago, rare mutations may be on the increase. The problem of missing heritability must also be due in part to the scope of studies. So far most of the subjects in genomewide studies have been European. As more of the world's genome is surveyed, the picture will inevitably become more detailed.

Some common disorders or traits may be explained by idiosyncrasies in the structure of the genome: Segments may be inverted or moved to different spots, and there are many varieties of repeats, like the CAG repeat in the huntingtin gene. New mutations underlie some disorders as well. A small percentage of cases of autism are caused by de novo point mutations, which is to say that while the mutations occur in genes, the condition isn't inherited. In addition to these dark-matter candidates, there is noncoding DNA. When geneticists discover that differences in DNA correlate with differences in health (for example, people with a certain condition have a T in one spot on the genome rather than an A), they have overwhelmingly found that these significant differences occur not in genes but in the noncoding regions of the genome. Why? They don't know.

It is also the case that the environment modulates genes, but which elements of the environment exactly? How well you slept as a child? How well you ate? The absence or presence of certain stressors matters too. Did you grow up in a war zone? Did your family live in poverty? Is anyone in your family an addict? What about your family history of disease and its level of education? Were you exposed to a large number of pollutants? Keep in mind that the way the environment shapes genes isn't through some vague influence: Everything we hear or see or feel or touch is translated into our tissue by the action of biochemicals of some kind, which should be traceable.

The lives that our parents and grandparents lived may also affect the way genetic conditions play out in our bodies. One of the central truths of twentieth-century genetics was that the genome is passed on from parents to child unaffected by the parents' lives. But it has been discovered in the last ten years that there are crucial exceptions to this rule. Epigenetics tells us that events in your grandfather's life may have tweaked your genes in particular ways. The classic epigenetics study showed that the DNA of certain adults in the Netherlands was irrevocably sculpted by the experience of their grandparents in a 1944 famine. In cases like this a marker that is not itself a gene is inherited and plays out via the genes. More recent studies have shown complex multigenerational effects. In one, mice were exposed to a traumatic event, which was accompanied by a particular odor. The offspring of the mice, and then their offspring, showed a greater reactivity to the odor than mice whose "grandparents" did not experience such conditioning. In 2014 the first ancient epigenome, from a four-thousand-year-old man from Greenland, was published. Shortly after that, drafts of the Neanderthal and Denisovan epigenomes were published. They may open up an entirely new way to compare and contrast our near-relatives and ancestors and to understand the way that they passed down experiences and predispositions. As yet it's unclear for how many generations these attachments to our genes might be passed down.

Even given our ability to read hundred of thousands of letters in the DNA of tens of thousands of people, it turns out that—at least for the moment—family history is still a better predictor of many health issues. For example, it is the presence of a BRCA mutation *plus* a family history of breast cancer that most significantly raises a woman's risk of the disease.

One of the most practical ideas to emerge out the intricate and complex activities of modern genetics is that we should think about our genes as risk factors. The presence of a BRCA mutation doesn't indicate for certain that you will develop breast cancer, but it increases your risk. There may be other factors in your life, not just genetic mutations but also family history and personal experiences, that compound the risk further. In the case of Huntington's disease, the presence of forty or more CAG repeats on the huntingtin gene currently means the risk of death from the disease is overwhelming. But when treatments that slow down the progression of the disease or even prevent its development are developed, the risk assessment will change.

Consider a less lethal but still important mutation on the F5 gene. People with this mutation will produce factor V Leiden (rather than the normal factor V), a protein that increases the likelihood of excessive blood clotting. If people with this mutation spent a great deal of time on airplanes, they would be especially vulnerable to the blood clots that occur in long-haul-flight syndrome. Armed with this information, they could be scrupulous about getting up and stretching frequently during a flight.

Until a few years ago, the only way for most people to access information about their own genomes was through genetic counseling. The service was offered before or during pregnancy when a genetic disorder was suspected, and for the most part the disorders that were detected were Mendelian, so the information was tremendously consequential. Genetic test results were presented in person by a professional who was trained to educate and assist. Since 2007, however, anyone has been able to send off a cheek swab or spit sample and learn about many of their genetic risk factors.

23andMe, for example, looked at genetic markers for Parkinson's disease, multiple sclerosis, and diabetes, as well as many other conditions. It also determines genetic susceptibility to certain drug reactions. I found out that I was likely to be more sensitive than most people to warfarin, a drug that is administered to prevent blood clots in emergency situations, like strokes. Should I need the drug at any point, doctors will ideally minimize the amount so that I don't bleed too much. I also learned that I was four times more likely than the average person to develop celiac disease. The knowledge made me pay attention to a vague set of gastric symptoms I'd had for more than a year. When I finally went to a doctor and did some testing, I found out that I did not have celiac disease, but I did have a food intolerance that has completely changed how I eat. The symptoms have now disappeared.

My husband waited for his 23andMe results with great anticipation, as his mother had died of multiple sclerosis when he was twenty-one. Although many people with multiple sclerosis live compromised but full lives, CB's mother was physically and mentally devastated to the point where she was unable to recognize her children. Because multiple sclerosis can be hereditary, it was the first condition CB checked when he received his genetic analysis. The report said that he was less likely to develop MS than the average person. That result doesn't mean he won't develop MS, but it does suggest that, as far as science can tell right now, he won't *inevitably* develop it.

Despite the obvious utility of such information, there is a serious debate in the medical and genetics communities about whether people should be allowed to access their own health-related genetic data. In 2013 the FDA suspended 23andMe's health service, and though the company and the regulatory body are now in discussions, it is unclear when the service will be restored. At issue is the accuracy of the information being offered and the belief that genetic information should have special safeguards placed upon it. This is quite clearly true for Mendelian diseases. What if consumers discovered terrifying news about themselves or family members? The risk is not to be taken lightly. But, as we now know, the majority of risk information to be gained from the genome is not Mendelian but relies on many factors.

Robert Green, a physician-scientist at Brigham and Women's Hospital and Harvard Medical School, told me that much of the fearful attitude toward genetic information was formed in the cauldron of genetic counseling. "Huntington's disease has been the paradigm for genetic testing for a long time. It's really not a very good paradigm, because there's a certainty to it, and most other genetic, even Mendelian genetic, variations are not fully penetrant."

As with almost every aspect of the practical use of genetics, there is currently more opinion than research into many basic questions. Is the discovery of a significant genetic risk by an individual outside the medical establishment as potentially harmful as many people fear? Preliminary studies have shown that what researchers suspect will be devastating news for people may not necessarily be so. Green used to work with the design and implementation of drug-testing clinical trials, so he decided to test the notion of treating information as if it were a drug: Will the information cause benefit or harm or both? The particular information shared in his study was whether subjects had a common variant of the APOE gene, which is connected to a high risk of Alzheimer's. "We designed the study carefully with many safety features and we initially did the study with the minimum number of people necessary to answer the question. The results of this study and several others which eventually evaluated over one thousand subjects was that among volunteers, disclosure of genetic risk information, even for a disease as frightening and untreatable as Alzheimer's disease, was quite safe," he said.

As far as direct-to-consumer testing results are concerned, there are relatively simple ways to filter information. 23andMe graded its results according to how confident it was of the science, and it shared those confidence levels with its customers. If the genetic risk factors for a serious condition were analyzed, customers had to specifically choose to "unlock" their results, so it was not possible to simply stumble upon them. (23andMe did not test for Huntington's disease.)

When I asked Jeff Carroll what he thought of direct-to-consumer genetic testing, his main concern was the unregulated testing of children. With Huntington's, if there is no question of early onset, it is considered profoundly unethical to test children, as a positive result would severely impact the way that child was treated and how he would feel about himself. Only a child himself, once he reaches the age of eighteen, is legally allowed to initiate such a test.

Even if people aren't harmed by learning about their personal genetic risks, critics wonder if they will use the information to improve their lives, to support the health of their families, and to reduce the enormous costs of health care in the developed world. Indeed, small studies by Navigenics and other companies have shown that there is not a lot of proactive response when people are informed about their genetic risks. In one study subjects were found to have a higher than normal risk for type 2 diabetes. It is well known that onset of the disease is affected by lifestyle, yet even when the at-risk subjects were given information about their susceptibility, many did not adjust their fat intake or increase exercise or consult medical specialists to minimize their risk.

The irony here is that this finding supports the argument that most genetic information is not exceptional and should not be treated as such by regulators. It's well known that any kind of health information—whether it is heart related, weight related, or aging related—often has little impact when it comes to getting people off the couch. Yet no one is suggesting that the medical establishment should stop trying to communicate information about exercise and food to the public, only that it needs find better ways to share it.

Do people even want to know their results? At a genetics conference in 2012, I listened to a presentation about a series of procedures for subjects who said yes to receiving information about whether they had a gene that was implicated in a certain type of cancer. The researchers found that although everyone initially opted in and made appointments to follow up and receive personal risk information, many didn't turn up to their first appointment to learn more about the condition, and fewer still turned up for the second, at which they would have received their results. This was interpreted as suggesting that people didn't want to know what their risk factor was. But what if the declining response was actually an indication that people are busy and that if institutions make it too hard to obtain information, subjects won't make an effort to pursue it? It may be not the fear of risk but the significant time commitment that's getting in the way of people's learning about preventable conditions.

One researcher surveyed members of a poor community and asked what their response might be to information about their genetic risks. Many were extremely fearful and superstitious, and one man told the researcher that he didn't want to know his results for fear that simply learning about them would somehow cause the condition to occur. Clearly, education is necessary to help many people understand that genes are almost never by themselves fate. One of the most direct ways to do this is to help people become acquainted with their own genome.

So far there has been little penetration of genomic data into the medical community. Customers of 23andMe and other direct-to-consumer companies describe taking their reports in to doctors who will often not even glance at the data. "How do clinicians cope with a genomic report? How should a report be designed for them? How will they cope with the targeted findings, and how will they cope with incidental findings?" asked Robert Green. All of these questions lag behind the rapid production of massive amounts of genomic data. "Why should genomic information be different than any other sensitive medical information a physician handles, like medical history, psychiatric history, substance abuse, HIV positivity, sexual orientation? Physicians are privy to all sorts of sensitive information and have responsibility for privacy and translation to their patient," Green observed. "Unless medicine is foolish and abdicates its responsibilities, it's really incumbent upon conventional medicine to sensibly integrate this into the practice of medicine."

The most important requirement for direct-to-consumer testing is that the information it provides about genetic risk be reliable. In a number of studies agencies of the U.S. government and scientists have compared the results of genetic reports from different testing companies, and in all cases there was divergence in how the companies analyzed the results. Some differences will be inevitable, but for certain disorders the same genome, depending on the company, was reported to be high-risk, low-risk, and no-risk. For critics that finding was damning enough to advise against any genetic testing.

Ultimately anyone who participates in personal genetic testing, whether it is historical or health related, should understand that these are thrilling, rich, but early days in this particular science. This moment is to the genomic future as the 1970s and early 1980s were to computers. Back then Steve Wozniak, Steve Jobs, and the unsung heroes of the digital age were tinkering in their garages with proto–personal computers. Now most people use not one or even two but many digital devices on a daily basis. If someone turned off all the computers tomorrow, the world would stop. People who use genetic services need to tolerate a degree of uncertainty, as advancements will be made. Yet there will always be some uncertainty where the genome is concerned.

———

In 1998 the American Anthropological Association issued a statement about race and physical variation that as of 2014 still stands:

> Historical research has shown that the idea of "race" has always carried more meanings than mere physical differences; indeed, physical variations in the human species have no meaning except the social ones that humans put on them.

It is remarkable that the idea that physical variations between groups have no meaning beyond the social still has considerable influence. We are all genetic creatures, and our families; small, medium, and large population histories; in-Africa and out-of-Africa experiences; and other massive historical contingencies shape the probabilities that play out in our lives.

The sharp edge of a genetic legacy is a fatal Mendelian disease. But this is just one of the ways that genes affect our lives. There are recessive diseases and complex disorders whose frequency is affected by ancestry. There are traits that shape our appearance, which people react to, which then shapes our reactions, feelings, and behavior. Being a creature that is formed from the genomes of two other creatures via a lottery of bits has many consequences, none of which can be completely understood without taking into account our DNA.

DNA tells us that we are creatures of chance and fate and that no one has quite the same mix of the two in his or her life. We think of ourselves as essentially whole, but when we look at our genome, we see that we are composed of many fragments stuck together. Many of our bits have different histories, and they each bring different probabilities into our lives. "Nobody is going to have a complement of alleles that is totally perfect," Jeff Carroll observed:

> There is a particularly Western tendency to think that we're pristine or untainted . . . that we can go off and do anything we want if we work hard enough and nothing can stop us except our own efforts. It's a really, really excellent lie for us all to believe in. But we need

to know it's a lie. . . . There are those of us that have extreme
cases . . . [but] even with one of these extreme cases you can still be
a useful human being and do good work.

We are also creatures of changeless truths and of interesting possibilities. Once you are born, your spot in the tree of humanity is fixed.
You will always have emerged out of everything that shaped the tree before you—the biology and the history. The millions of bits that initially
made you—all the cultural bits and the genetic bits, each with its risk
factors, predispositions, and probabilities—were shaped by that past.

As you develop and grow older in whatever world you live in, the calculations change. Your family, the history of your community, your government, and even your food alter them. *You* alter them. Why not find
out what the calculations are? Doing so will not guarantee that you will
learn what *will* happen, but it may help you to consider the possibilities.
Your genome is just the first hand that life deals you. How you play it is
up to you.

Epilogue

During the time I've been writing this book I've met many people who have lost someone, not in the poignant, fading way that older people eventually leave us but with a sense of sudden or unnatural separation. A friend came to stay with us and told us about his father, who was born in Taiwan. He was adopted as a young boy but only discovered that fact as an adult, when his wife-to-be's family investigated his background. It fell to his wife to tell him that the people he thought were his parents were not. Later he searched for his parents and found a woman in a small village who claimed him as her own. He sent her money and kept her comfortable until she died. His son now suspects that she wasn't a relation at all. Our friend recently sent his DNA to two genetic-genealogy companies, expecting to be told that his ancestry was entirely Chinese, but both reported that 7 percent of his DNA likely came from somewhere else. One of them suggested the mysterious 7 percent was Polynesian.

A woman who works in the same building as I do told me that her father was left as a baby on a doorstep in England. All she knew of his parents was the name of his mother. She started to dig through records, and they sent his DNA to a genetic-genealogy company. She found a handful of women with the same name, and she narrowed the candidates to one. Every time I see her, another small step has been taken.

I met a geneticist whose parents were particularly old, something he had noticed as a child but had never really thought about. Only as an adult did he discover that his "sister" was not in fact his sister but his mother. He was raised by his grandparents.

I haven't yet found out much about my own paternal grandfather. Since I began this book, my parents and I have slowly begun to speak about genealogy, and while it has been hard for them they have supported my desire to know. The search goes on.

One of the most interesting findings of twenty-first-century genetics is that it is not as determinative as we feared. Increasingly, however, we find evidence that cultural history shapes us more than we knew. How long can we expect the effects of culture to persist? Karla Hoff, a senior research economist at the World Bank, provides insight through her examination of caste divisions and decision making in India.

The caste system is a form of social division that dates back thousands of years. While it is thought that historically there was some flexibility for individuals to change their caste, the system has been rigidly stratified since British rule in India. Higher castes have always had more freedom, greater status, and more rights. Typically, low castes were allowed to take only jobs that no one else wanted to do, like working with corpses or cleaning toilets. High castes called the lowest of the low "untouchables."

Untouchability was outlawed fifty years ago, and today legal differences among castes have been eradicated. Many social differences have flattened out too. Just because someone belongs to a high caste doesn't mean he is necessarily wealthy, highly educated, or connected to his local political scene. Nevertheless social boundaries among castes continue to be vigorously policed by the community. It's not uncommon for married couples with one high-caste spouse and one low-caste spouse to be lynched, raped, or beaten. In April 2014 a seventeen-year-old Dalit boy was beaten and strangled by a group of high-caste men because they saw him sitting in a field with their sister. Discriminatory practices and caste-based violence are still reported in 80 percent of Indian villages.

Hoff investigated how modern-day levels of group solidarity have been shaped by the caste system. She set up a number of experiments in which three individuals from either high or low castes played a game that involved cheating and punishment. She found that people in low castes were disinclined to punish anyone, regardless of which caste the cheater belonged to. People in high castes, however, were happy to punish cheaters from other castes who hurt members of their own caste.

Even when wealth, education level, and participation in politics were accounted for, the caste effect was the same: The already-privileged had one another's backs.

Hoff believed that a culture of righteous punishment and a strong group identity were passed down in the high-caste families, helping them maintain their power. Recall that high castes have had power in India for an extraordinarily long period of time, so it's likely that these attitudes have been passed down by thousands of family groups over many centuries.

The question for a development economist like Hoff is whether these incredibly ancient attitudes can be changed. She is hopeful and described a recent study that tracked social change in India after a 1993 constitutional amendment required that in one third of all villages, the position of village chief must be reserved for women. "What they found is that in the course of just seven years, being exposed to village leaders who were women eliminated the bias men had against women leaders in evaluating them," Hoff said. "Men still didn't like women leaders but they didn't evaluate their performance as leaders unfairly. Parents' aspirations for their daughters went up and the level of domestic violence that was reported went way up, even though the village leaders have no jurisdiction over the police. Even after the reservations ended, women ran for office and were more likely than before to win elections. The results suggest the possibility of a sea change in culture."

"If you believe that culture was a set of rules of thumb—don't trust a woman, don't elect a woman, don't educate a daughter—that were very resistant to change because it was passed down in your mother's milk," Hoff said, "then you couldn't make sense of this."

The tools we now have for making sense of cultural history and the personal past are nothing less than extraordinary. If everyone had his DNA analyzed, for example, and that information were linked to everyone's historical information, it would be the nearest thing to the book of humanity. But *should* we get our DNA analyzed? There are practical considerations and potentially negative consequences. There are insurance issues, privacy issues, health issues, and questions about how the whole industry will develop. Anyone who wants to have his DNA analyzed should consider all of these factors.

In 2010 I had my DNA analyzed by the Icelandic company de-CODEme. The organization was the first to offer an in-depth personal DNA analysis over the Web. Its parent company, deCODE, is a famous innovator in genomic research. In 2013 deCODE was taken over by one of the world's biggest biopharmaceutical companies, which is publicly traded. Presumably the company that acquired this asset must answer to its investors about how to best profit from it. At the time deCODEme stopped taking new customers.

When the company was sold, I did not receive information about de-CODEme's suspension or deCODE's new owners. As of May 2014 I was still able to log in to see my deCODEme analysis. But it does not appear to have been updated at all. What will happen in a few years, in ten years, in one hundred? It made me wonder about what policies companies like this would follow if they changed hands. Different companies make different commitments to their users, and it is sensible to read the fine print before sending in a sample. But also consider that no matter what is said now, while any company can be shut down, your DNA sample may live on and it will point to you forever.

George Church runs Harvard's Personal Genome Project, the goal of which is to amass a collection of complete genomes and make them—and their owners—available to scientists for research. Now any individual may volunteer his own genome to the project, and if he is accepted, he will pay to have it sequenced. Each year many PGP members attend a special conference in Boston with scientists who are eager to explore their genomes and the traits they may underlie. At the 2013 conference I saw PGPers undergo smell tests, submit bacterial swabs, and guess at the identity of well-known people. Everyone seemed to be having a great time.

Church kicked off the project with just ten volunteers. Cognitive psychologist Steven Pinker was PGP number 6. His genome is now available on the Internet, along with his medical history. "With the genome no less than the Internet," he wrote at the time, "information wants to be free." In 2014, he described the FDA intervention in 23andMe as regulatory overreach. "No harm can come from people knowing more about their genomes, and it's none of the government's business if they want to do so," he said. PGP pioneer number 4, Misha Angrist, a professor at Duke University, told me the sense of anticipation he had that sequencing his genome was going to change his perception of himself was much greater than the reality. Now Angrist checks his genome every few months. "One thing I notice is alleles that were identified as high risk have been downgraded," he said. "That is to be expected. As we sequence more and more genomes, we start to see these mutations that we might have thought, *Well, it looks like it really messes up a protein*, and in fact there are thousands or millions of us walking around with them." He added, "It's a reminder of how naive we are, how simple our understanding of human biology is."

When I met PGP contributor number 3, investor and technology commentator Esther Dyson, she told me that when she had chicken pox as a child she carefully counted the spots on her face every day. Even then, she said, she was curious about the "quantified self." What had she learned about herself from her quantified genome? In fact, it was not a great deal, she said, and this was true for most people. "It's like learning American history, you may not learn a lot about yourself individually but you learn a lot of context," she explained. "To be an educated person, this is the sort of stuff you should know . . . You will understand things better."

As an afterthought, she added, "Actually, I have one APOE4, so I have double the normal chance of getting Alzheimer's. But honestly it hasn't changed my life; it hasn't changed how I think about things. I'm well aware I'm going to die someday, and everything else really is detail."

Church told me that early in the development stage of the project it became clear that it would be impossible to guarantee the privacies that are traditionally given to experimental subjects. DNA is an inescapable identifier, and we live in an age when it is virtually impossible to retract information from the public domain once it has been shared. Indeed, a 2013 study showed that using DNA data from yet another database, the 1000 Genomes Project, along with the age and location data that the

project made publicly available, researchers were able to use a genealogy Web site to identify by name five people they chose from the 1000
Genomes set. Because they found the individuals' family trees, they
found their families too.

Church's approach has been to pioneer "open consent," which means
that there is no promise of anonymity. Volunteers are counseled and vet-
ted; they must prove they understand what is at stake before they are
allowed to join.

Many people I spoke to were worried that insurance companies might
obtain their genetic data and, despite the fact that most DNA is not determinative, adjust their policies in light of it. There is cause for concern.
Currently in the United States the Genetic Information Nondiscrimina-
tion Act prevents health insurers from forcing customers to share genetic
information, although they may request that an individual take a particular test if there is a known risk. However GINA does not apply to life,
disability, and long-term-care insurance. In Britain a moratorium on
making the results of DNA tests available to insurers has been extended
a number of times and is currently set to expire in 2017. British insurers,
however, require that when a life insurance policy over £500,000 is issued, test results for Huntington's disease be disclosed. In Australia individuals who have taken genetic tests may be forced to give the results
to insurance companies. In 2013 one company even offered its customers a discounted genetic assay. The offer was pitched as a positive way to
help people look after their health, but the fine print stated that people
might have to disclose the test results. Currently many European countries ban genetic discrimination, although Canada has no law against it.

Now governments are carrying out large-scale genomewide association studies to investigate common diseases and rare mutations, and
the biological networking of populations will only increase. In addition
to projects in the United States, the United Kingdom, and many European countries, the government of the small Faroe Islands in the North
Sea, settled by Vikings in the ninth century, is planning to offer sequencing to its entire population of fifty thousand people. There are
extraordinary gains to be made, but if careful regulation is not put in
place, one wonders how insurance and pharmaceutical companies will

treat our grandchildren if they have genetic information about them, perhaps even before they are even conceived.

Insurance aside, in the coming years we will all be reading more stories about genetics and health as the gaps in missing heredity are filled in. It's important to keep in mind that researchers should be able to account for socioeconomic effects as well as genetic ones. Even now a number of studies have shown that what appears to be a genetic predisposition to a disease—say, type 2 diabetes—in a certain population may actually be a result either of that group's socioeconomic situation or of a combination of genetics and socioeconomics.

As researchers advance their understanding of culture and DNA, perhaps the most exciting prospect is a larger synthesized body of knowledge that explains the way that history affects DNA and the way that DNA affects history, with both together acting on some version of us, whether our private self, our family self, or our admixed self. A recent look at the folktales of Europe used the techniques of population genetics to study their distribution. The researchers found that stories cluster geographically in the way that genes do and that you can determine language and ethnic boundaries by the different versions of a tale. In fact, the effects of geography and language boundaries were stronger for stories than for genes, meaning that genes overlapped between groups in a more fine-grained way. As far as mixing goes, it looks as if it is typically easier for two people to mix their genes without many words in common than it is for a story to be passed through a language barrier. Or, as Razib Khan, a geneticist and science blogger, put it, culture is chunky, whereas genes are creamy.

Now that we can see so much more clearly what has been passed down to us, what will we choose to hold back, what will we send further down the line, and how will we pass that information on? We will do so in all the traditional ways, no doubt, but we also have one more method that is completely new. Noting that DNA is an extraordinary medium for the storage of high-density information, scientists have begun to explore its potential for use as a digital storage device. The first attempts to encode a trivial message with a few words in it took place at the end of the twentieth century; a decade later scientists were able to lay down complex stories in DNA, including Martin Luther King's "I have a dream" speech, a scientific paper, and, of course, Shakespeare's sonnets. Rather ingeniously this new technique may yet add more layers to the palimpsest of DNA.

ACKNOWLEDGMENTS

With admiration, thank you to Misha Angrist, Andrew Appleby, Tony Arthur, Gil Atzmon-Druze, Holly Choon Bachman, Barbara Barandun, Nola Beagley, Alan Bittles, Blaine Bettinger, Cinnamon Bloss, Sir Walter Bodmer, Baiying Borjigin, Jeff Carroll, Stanley Chang, George Church, Anna DiRienzo, Peter Donnelly, Eric Durand, Esther Dyson, Eran Elhaik, Eric Ehrenreich, Jim Ericson, Yaniv Erlich, Edward Farmer, Ellen Gunnarsdóttir, Marc Feldman, Cassandra Findlay, Jill Gaeiski, Ivy Getchell, Leanne Goss, Robert C. Green, Bennett Greenspan, Colin Groves, Helen Harris, John Hawks, Brenna Henn, Karla Hoff, Evan Imber-Black, Dan Jones, Turi King, Damian Labuda, David Allen Lambert, Andrei Lankov, Stephen Leslie, Donald MacLaren, Joe Mauch, Janet McCalman, Gavan McCarthy, Michael McCormick, Rhonda McClure, Glynis McHargue Patterson, Robert McLaren, Garry McLoughlin, Geoff Meyer, CeCe Moore, Joanna Mountain, David Murray, Leo Myers, Paul Nauta, David Noakes, Robert Noel, Nathan Nunn, Paul Nurse, Katy Oh, Peter Pan, Nick Patterson, Steven Pinker, Ugo Perego, Peter Ralph, David Reich, Mark Robinson, Thomas Robinson, Wendy Roth, Jacqueline Ross, Kevin Schurer, Richard Scott, Leonie Sheedy, Guido Tabellini, Shelly Tardashian, Jay Verkler, Nico Voightländer, Jennifer Wagner, Leonard Wantchekon, Bruce Whinney, Sloan Williams, Wayne Winkler, and Scott Woodward.

Special thanks to Gisela Heidenreich, Gudrun Sarkar, and Wolfgang Gliebe.

Thank you to everyone who spoke to me off the record and without attribution.

Thank you to Debra Hine, without whom this would have taken years longer, and to Daniela Diedrich. Good luck with your PhD! Thank you, Eric Maisel.

Thank you, Razib Khan.

Thanks Gavan McCarthy for the Heinlein quote. Thank you, Helen Harris, for your brilliant expertise and special thanks to Alison Alexander for a delightful afternoon.

Thank you to the Abbotsford Convent and all who work there.

Thank you lovely Stephen Armstrong for an opportunity that made me think above and beyond.

Thank you to my far-flung writers groups, including Simon Caterson and Cordelia Fine. Thank you, Anne Baker and John Katinos, for your warm friendship and accommodation. You too, PP. Special thanks to Sheri Fink, Susan Cain, Monica Dux, Caleb Crain, Peter Terzian, Libba Bray, and Marci Alboher.

Amanda Schaffer, you are an inspiration, and I owe you one!

Shelagh Lloyd, you are kind and brilliant.

Thank you so much to everyone at Viking and William Morris Endeavor, including Francesca Belanger, Hilary Roberts, Nicholas Bromley, Hal Fessenden, Shannon Twomey, and especially to my editor, Rick Kot, and my agent, Jay Mandel.

Some of the words in the book have previously appeared in articles I wrote for the *Sunday Age*, the *Good Weekend*, the *Monthly*, *MIT Technology Review*, NewYorker.com, and other publications. My thanks to John van Tiggelen, Ben Naparstek, Mary-Anne Toy, Brian Bergstein, Jay Kang, and the other magazine editors who have engaged with this work.

Thanks as always, Nessie, for lending me your books.

Thanks for sharing your tree Bob and Eileen Jukes.

Thank you Damien Kenneally and Mary Kenneally.

Thank you Conrad Mackle, Angela Kenneally, Michael Jukes, and Allen Baldwin.

With deep gratitude to my mother and father and to Hugh, Katherine, Steve, Angie, Mick, Shelagh and Simon.

Three billion bases of thanks and love to Chris Baldwin, Nat, and Fin.

NOTES

Author's Note

ix **"blurred, if not dissolved"**: McCormick further explains his approach in Jonathan Shaw, "Who Killed the Men of England?" *Harvard Magazine*, July–August 2009, available at http://harvardmagazine.com/2009/07/who-killed-the-men-england.

Introduction

12 **the QWERTY design**: Jared Diamond popularized the problem of QWERTY, as he has many other issues of path-dependent history.

Chapter 1: Do Not Ask What Gets Passed Down

20 **Genealogy, wrote *Guardian* columnist Zoe Williams**: Z. Williams, "Ancestor Worship," *Guardian*, November 8, 2006, available at http://www.theguardian.com/commentisfree/2006/nov/08/comment.zoewilliams.

20 **"Show me a genealogist"**: S. Sanghera, "Every family has a story—but don't tell me," *Times*, April 7, 2010, available at http://www.thetimes.co.uk/tto/opinion/columnists/sathnamsanghera/article2470220.ece.

20 ***Genealogy is Bunk:*** R. Conniff, "Genealogy is Bunk," *Strange Behaviors*, July 1, 2007, available at http://strangebehaviors.wordpress.com/2007/07/01/genealogy-is-bunk/.

20 **Richard Conniff**: Despite his reservations, Conniff also acknowledged some of the pleasures of family history. He wrote that he would like to know more about his family name, which in Irish means, "son of a black hound." He also confessed to a deep curiosity about his "Italian great-grandfather, who used to chase his father down Webster Avenue in the Bronx swinging a sickle and yelling, 'I catch-a you, I keel-a you.'"

23 **the Reverend Ephraim Newton**: R. Lewontin, "Is There a Jewish Gene?" *New York Review of Books*, December 6, 2012, 59 (19).

24 **"All the laws of Washington"**: A. Clymer, "Strom Thurmond, Foe of Integration, Dies at 100," *New York Times*, June 27, 2003, available at http://www.nytimes.com/2003/06/27/us/strom-thurmond-foe-of-integration-dies-at-100.html.

25 **"I was always the family record keeper"**: Quotes from Wendy Roth in this chapter are from my interviews with her.

28 **In 2012 Jordi Quoidbach:** J. Quoidbach, D. T. Gilbert, and T. D. Wilson, "The End of History Illusion," *Science* 339, no. 6115 (2013): 96–98.

30 **the novelist Will Self:** E. Day, "Will Self: I Don't Write for Readers," *Guardian*, August 5, 2012, available at http://www.theguardian.com/books/2012/aug/05/will-self-umbrella-booker-interview.

30 **A famous study compared:** J. Henrich, S. J. Heine, and A. Norenzayan, "The Weirdest People in the World?" *Behavioral and Brain Sciences* 33, no. 2–3 (2010): 61–83.

Chapter 2: The History of Family History

34 **"not genealogists constitutionally":** F. Weil, *Family Trees: A History of Genealogy in America* (Cambridge, MA: Harvard University Press, 2013). I was greatly influenced by Weil's book. It was a wonderful source as well as a window onto a body of work it would have otherwise taken me years to track down. Many of the examples in this chapter of the development of genealogy in America either came from it or pointed me toward fruitful further research.

34 **"But they are not interested in genealogy":** Quotes from David Allen Lambert in this chapter are from my interviews with him.

36 **Many modern genealogies:** We don't know how the much larger proportion of the population that wasn't literate thought about family or tracked heritage at this time, but if they didn't think about it at all, they would be one of the only groups who didn't.

37 **bourgeoisie adopted the practice:** C. Klapisch-Zuber, "The Genesis of the Family Tree," *I Tatti Studies in the Italian Renaissance* 4 (1991): 105–29.

39 **In an 1815 letter:** F. Weil, *Family Trees: A History of Genealogy in America* (Cambridge, MA: Harvard University Press, 2013), 42.

39 **"product of tangled impulses":** F. Weil, *Family Trees: A History of Genealogy in America* (Cambridge, MA: Harvard University Press, 2013), 4.

40 **Michelle Obama's family history:** Initially traced up to five generations back on all branches by genealogist Megan Smolyenak.

40 **"arguably the element":** F. Weil, *Family Trees: A History of Genealogy in America* (Cambridge, MA: Harvard University Press, 2013), 2.

41 **"These questions are not for publick information":** F. Weil, *Family Trees: A History of Genealogy in America* (Cambridge, MA: Harvard University Press, 2013), 42.

42 **"Our age is retrospective":** R. W. Emerson, *Nature*, 1856, available at: http://oregonstate.edu/instruct/phl302/texts/emerson/nature-emerson-a.html.

42 **"When I talk with a genealogist":** F. Weil, *Family Trees: A History of Genealogy in America* (Cambridge, MA: Harvard University Press, 2013), 47.

43 **"excessively aristocratical":** F. Weil, *Family Trees: A History of Genealogy in America* (Cambridge, MA: Harvard University Press, 2013), 81.

45 **"witnessed the emergence of the first generation":** F. Weil, *Family Trees: A History of Genealogy in America* (Cambridge, MA: Harvard University Press, 2013), 158.

46 "Daniel Webster": F. Weil, *Family Trees: A History of Genealogy in America* (Cambridge, MA: Harvard University Press, 2013), 49.

Chapter 3: The Worst Idea in History

50 make a "stamp" on offspring: R. J. Wood, "The Sheep Breeders' View of Heredity (1723–1843)," in *Proceedings of the 2nd Conference: A Cultural History of Heredity* (Berlin: Max Plank Institute, 2003).

51 "Flying squarely in the face": *Breeders Gazette*, available at http://www.ans .iastate.edu/history/faculty/bakewell/bakewell.html.

51 psychological traits being passed down: C. López Beltrán, "Heredity Old and New: French Physicians and L'hérédité naturelle in Early 19th Century," in *A Cultural History of Heredity II: 18th and 19th Centuries* (Berlin: Max -Planck-Institute for the History of Science, 2003), 7–19, available at http:// search.wellcomelibrary.org/iii/mobile/record/C__Rb1748331__Sheredity__ P0,10__Orightresult__X6.

55 eugenics could be a new religion: F. Galton, "Eugenics: Its Definition, Scope, and Aims," *American Journal of Sociology* 10, no. 1 (1904): 1–25.

56 moments in Grant's intellectual development: J. Spiro, *Defending the Master Race: Conservation, Eugenics, and the Legacy of Madison Grant* (Burlington, VT: University Press of New England, Hanover and London, 2009).

57 "The African Pygmy": J. Spiro, *Defending the Master Race: Conservation, Eugenics, and the Legacy of Madison Grant* (Burlington, VT: University Press of New England, Hanover and London, 2009), 46.

58 "The immigrant laborers": M. Grant, *The Passing of the Great Race* (New York: Charles Scribner's Sons, 1936 4th ed.), available at https://archive .org/stream/passingofgreatra00granuoft/passingofgreatra00granuoft_djvu.txt.

59 "This is a practical, merciful and inevitable solution": M. Grant, *The Passing of the Great Race* (New York: Charles Scribner's Sons, 1936 4th ed.), available at https://archive.org/stream/passingofgreatra00granuoft/passingof greatra00granuoft_djvu.txt.

60 "The cross between a white man": M. Grant, *The Passing of the Great Race* (New York: Charles Scribner's Sons, 1936 4th ed.), available at https://archive .org/stream/passingofgreatra00granuoft/passingofgreatra00granuoft _djvu.txt.

60 *The Passing of the Great Race* was his Bible: I found this story mentioned many times in articles and books but could not locate a definitively original source.

61 "the most valuable classes": M. Grant, *The Passing of the Great Race*, (New York: Charles Scribner's Sons, 1936 4th ed.), available at https:// archive.org/stream/passingofgreatra00granuoft/passingofgreatra00granuoft _djvu.txt.

61 "Pure + Pure": Marriages-Fit and Unfit photograph, ID 11508, Cold Spring Harbor Laboratory DNA Learning Center, available at http://www.dnalc.org/ view/11508--Marriages-Fit-and-Unfit-.html.

62 **the transmission of "defective strains"**: *American Breeders Magazine* 3 (1912; reprint, London: Forgotten Books, 2013).

63 **"determine the fate of our society"**: Unless otherwise noted, all Popenoe quotes are from P. Popenoe and R. H. Johnson, *Applied Eugenics* (New York: Macmillan, 1918), p 341–49.

63 **A 1925 book reviewer:** "Whither Marriage?" *New York Times*, April 19, 1930, p. 85.

63 **"the sacred thread of immortality"**: P. Popenoe and R. H. Johnson, *Applied Eugenics* (New York: Macmillan, 1918), 351, available at http://hdl.handle.net/1805/1042.

64 **"From an historical point of view"**: P. Popenoe and R. H. Johnson, *Applied Eugenics* (New York: Macmillan, 1918), 184, available at http://hdl.handle.net/1805/1043.

65 **before contracts were signed:** Such detective agencies had been around since the 1890s. My main source for information on Japanese eugenics was J. Robertson, "Blood Talks: Eugenic Modernity and the Creation of New Japanese," *History and Anthropology* 13, no. 3 (2002): 191–216.

Chapter 4: The Reich Genealogical Authority

67 **"family in the service of *Rassen*"***:* Quotes from Joe Mauch in this chapter are from my interviews with him.

69 **genealogists had significant social influence:** E. Ehrenreich, *The Nazi Ancestral Proof: Genealogy, Racial Science, and the Final Solution* (Bloomington, IN: Indiana University Press, 2007), pp. xx and 234. Much of my history of genealogy in Nazi Germany comes from personal communications with Ehrenreich and from his book.

71 **"effects of racial crossing"**: This analysis and the following, including the observation about Darwin's finches, come from V. Lipphardt, "Isolates and Crosses in Human Population Genetics; or, A Contextualization of German Race Science," *Current Anthropology* 53, no. S5 (2012): S69–S82.

71 **"The line between promoting the idea"**: E. Ehrenreich, *The Nazi Ancestral Proof: Genealogy, Racial Science, and the Final Solution* (Bloomington, IN: Indiana University Press, 2007), 48.

72 **"by the time the Nazis assumed power"**: E. Ehrenreich, *The Nazi Ancestral Proof: Genealogy, Racial Science, and the Final Solution* (Bloomington, IN: Indiana University Press, 2007), 45.

72 **"Dogs and horses"**: E. Ehrenreich, *The Nazi Ancestral Proof: Genealogy, Racial Science, and the Final Solution* (Bloomington, IN: Indiana University Press, 2007), 134.

72 **"virtually dependent on one's family chart"**: B. Gausemeier, "Genealogy and Human Heredity in Germany, Late 19th and Early 20th Centuries," 2011, available at http://wwwold.mpiwg-berlin.mpg.de/de/forschung/projects/DeptIII-BerndGausemeier-GenealogyAndHumanHeredity.

73 "**While other branches of learning**": E. Ehrenreich, *The Nazi Ancestral Proof: Genealogy, Racial Science, and the Final Solution* (Bloomington, IN: Indiana University Press, 2007), 136.

74 "**only one most holy human right**": Translated by Eric Ehrenreich and Daniela Diedrich. Ehrenreich showed me photographs of an Ahnenpass cover and the dedication on this inside page.

74 "**No one knows my unbelievably heavy sorrow**": E. Ehrenreich, *The Nazi Ancestral Proof: Genealogy, Racial Science, and the Final Solution* (Bloomington, IN: Indiana University Press, 2007), 113.

76 "**You know, as a little child**": Quotes from Gisela Heidenreich in this chapter are from my interviews with her.

80 "**My eyes aren't perfect**": M. Landler, "Results of Secret Nazi Breeding Program: Ordinary Folks," *New York Times*, November 7, 2006.

Chapter 5: Silence

85 "**We weren't allowed to talk to each other**": Quotes from Geoff Meyer in this chapter are from my interviews with him.

88 **One ex-ward**: The Care Leavers Australia Network has a newsletter in which the classifieds read like ads for former selves: "Michael would like to get in touch with anyone that remembers him in Renwick in the 1950s"; "If anyone can remember my nickname Debbie Wobble Head from the Ballarat Children's Home, it would be nice to have some contact."

89 **access to this fundamental information**: It may be that people born in the Federal Witness Protection Program are also denied access to their original birth certificate. It's unclear to how many individuals this has applied.

90 "**Ivy, my little mate**": Getchell did see her father again before he died, but he didn't tell her that he had tried to find her, so she never knew. Perhaps he thought she received his letter but did not want to respond.

93 **family connection to a dissident**: My thanks to Katy Oh and Andrei Lankov for this information.

93 **fallout from the recent dictator**: Biographical information about Baiying Borjigin comes from my interview with him and from B. Borjigin, *Searching for My Source: A Descendant of Genghis Khan* (Canberra, Australia: Australian Chinese Culture Exchange and Promotion Association, 2010).

95 "**The meals and bodies**": B. Borjigin, *Searching for My Source: A Descendant of Genghis Khan* (Canberra, Australia: Australian Chinese Culture Exchange and Promotion Association, 2010), 18.

95 "**If you can restore this list**": B. Borjigin, *Searching for My Source: A Descendant of Genghis Khan* (Canberra, Australia: Australian Chinese Culture Exchange and Promotion Association, 2010), 24.

96 **find my family's origins**: This historian was Helen Harris.

98 **They were often hungry too**: This was because other passengers stole their food, not because there wasn't enough in the first place.

99 **"Never mind, dear"**: Unless otherwise cited, quotes from Alison Alexander in this chapter are from my interviews with her.

99 **she know that was the case?**: Much of the information about Tasmanian convicts in this chapter comes from my interviews with Alexander and from her book, A. Alexander, *Tasmania's Convicts: How Felons Built a Free Society* (Crows Nest: Allen & Unwin, 2010). In many cases I came across the most interesting stories, statistics, and quotes through her first.

100 **that was wicked in another**: J. Braithwaite, "Crime in a Convict Republic," *Modern Law Review* 64, no. 1 (2001): 11–50.

100 **shipload of convicts in 1812**: The English sent convicts to the United States starting in the seventeenth century. After 1829 more than two and a quarter million convicts were transported from one country in the world to another, according to Braithwaite.

100 **wives and children sent from England**: Because the system assigned convicts to households, some were even assigned to carry out their sentences under their wives.

101 **"a new and splendid country"**: J. Braithwaite, "Crime in a Convict Republic," *Modern Law Review* 64, no. 1 (2001): 20-21.

102 **"sink of wickedness," . . . "den of thieves"**: A. Alexander, *Tasmania's Convicts: How Felons Built a Free Society* (Crows Nest: Allen & Unwin, 2010), 188.

102 **a society of dangerous people**: The notion that convicts were a blight was not confined to England. Indeed, the Australian penal colony was established because the United States refused to take any more of Britain's convicts. It is interesting, notes Alexander, that modern Americans have a degree of amnesia, or at least a distinct lack of interest, in their own convict past. For all of Benjamin Franklin's egalitarian bonhomie, he was not enthusiastic about lawbreakers. He set the tone when writing of British transportation: "Emptying their jails into our settlements is an insult and contempt, the cruelest that even one people offered to another." (A. Alexander, *Tasmania's Convicts: How Felons Built a Free Society* [Crows Nest: Allen & Unwin, 2010], 186.) Later, and more succinctly, he declared: "Send them back rattlesnakes!" (J. Braithwaite, "Crime in a Convict Republic," *Modern Law Review* 64, no. 1 [2001]: 7.)

102 **titillated by reports of bestiality and cannibalism**: Actually, there was one rather unpleasant people-eating incident. As Alexander tells it, in 1822 about half a dozen men ran away from a penal settlement and ended up starving in the bush until one of their number, the Irishman Alexander Pearce, killed and ate the others one by one. Pearce was rearrested and sent back to the settlement, and yet he escaped once more with a new companion, young Thomas Cox. Why did Cox run with Pearce, the alleged cannibal? We'll never know. Pearce ate him too. Pearce was caught again and finally hung in 1824.

102 **"They misquoted Latin"**: A. Alexander, *Tasmania's Convicts: How Felons Built a Free Society* (Crows Nest: Allen & Unwin, 2010), 204.

105 **"They melt from the earth"**: A. Alexander, *Tasmania's Convicts: How Felons Built a Free Society* (Crows Nest: Allen & Unwin, 2010), 217.

106 **it was the subject of relatively little research:** J. Crowley, W. J. Smyth, and M. Murphy. *Atlas of the Great Irish Famine* (New York: New York University Press, 2012), viii.

106 **hardly anyone spoke of it:** The pattern is reminiscent of François Weil, a Frenchman, being the first to write a book about the history of American genealogy. Overall, Irish culture left the stories of the famine inside the box of folklore and actively downplayed its significance.

Chapter 6: Information

112 **analog documents they held:** A. Shoumatoff, *The Mountain of Names: A History of the Human Family* (New York: Kodansha International, 1995), pp. xxi and 318.

114 **"The core concept":** Quotes from Jay Verkler in this chapter are from my interviews with him.

115 **he associates with personal strength:** M. P. Duke, A. Lazarus, and R. Fivush, "Knowledge of Family History as a Clinically Useful Index of Psychological Well-being and Prognosis: A Brief Report," *Psychotherapy: Theory, Research, Practice, Training* 45, no. 2 (2008): 268; M. P. Duke, "The Stories That Bind Us: What Are the Twenty Questions?" Huffington Post, March 23, 2013, available at http://www.huffingtonpost.com/marshall-p-duke/the-stories -that-bind-us-_b_2918975.html.

120 **they were impossible to read:** G. Palsson, "The Life of Family Trees and the Book of Icelanders," *Medical Anthropology* 21, no. 3–4, (2002): 337–67.

120 **"If *well housed*, the tiles would last ten thousand years":** Quotes from Gavan McCarthy in this chapter are from my interviews with him.

124 **"That's the classic genealogist":** Quotes from Dan Jones in this chapter are from my interviews with him.

127 **94 percent of all our stored information:** S. Wu, "How Much Information is There in the World?" Phys.org, February 10, 2011, available at http://phys .org/news/2011-02-world-scientists-total-technological-capacity.html#jCp, and M. Hilbert, "The World's Technological Capacity to Store, Communicate, and Compute Information," *Science* 332, no. 6025, 60-65.

129 **"When you have a hundred-percent count data":** Quotes from Kevin Schurer in this chapter are from my interview with him.

129 **demographics, longevity, and fertility:** H. Ledford, "Genome Hacker Uncovers Largest-Ever Family Tree," *Nature*, October 28, 2013, available at http://www.nature.com/news/genome-hacker-uncovers-largest-ever-family -tree-1.14037. The tree was built by Yaniv Erlich, whose light map project is also described.

130 **charges for online access:** In some special cases it is free.

132 **"a puzzle the size of a football stadium":** G. Pálsson, *Anthropology and the New Genetics* (Cambridge, UK: Cambridge University Press, 2007), 71.

133 **"Now a company in Iceland":** G. Pálsson, *Anthropology and the New Genetics* (Cambridge, UK: Cambridge University Press, 2007), 138.

133 "Accidentally sleeping with a relative": T. Sykes, "Iceland's Incest-Prevention App Gets People to Bump Their Phones Before Bumping in Bed," *Daily Beast*, April 23, 2014, available at http://www.thedailybeast.com/articles/2013/04/23/iceland-s-incest-prevention-app-gets-people-to-bump-their-phones-before-bumping-in-bed.html.

134 eighty-eight thousand marriages: H. Gauvin, et al., "Genome-wide Patterns of Identity-by-Descent Sharing in the French Canadian Founder Population," *European Journal of Human Genetics* 22 (2014): 814–21.

134 the population boom in all of Quebec was "spectacular": Unless otherwise cited, quotes from Labuda in this chapter are from my interview with him.

136 "the exceptional people": Quotes from Janet McCalman in this chapter are from my interviews with her.

136 "My great-grandfather": Quotes from Garry McLoughlin in this chapter are from my interviews with him.

137 "Some of them weren't nice": Quotes from Leanne Goss are from my interview with her.

138 "I hardly read novels anymore": Quotes from David Noakes are from my interview with him.

Chapter 7: Ideas and Feelings

139 With no time to scream: My information about Equiano's life and the plight of Africans during the slave trade came from many sources, including the writings of Nathan Nunn and Leonard Wantchekon, as well as Olaudah Equiano's book: O. Equiano, *The Interesting Narrative of the Life of Olaudah Equiano (Written by Himself)* (Project Gutenberg EBook, 2005).

140 "red water ordeal": W. Hawthorne, "The Production of Slaves Where There Was No State," *Slavery and Abolition* 20, no. 2 (1999): 97-124, via N. Nunn, "The Long-term Effects of Africa's Slave Trades," *The Quarterly Journal of Economics* 123, no. 1 (2008): 139–176.

141 sell them to slavers: C. Piot, "Of Slaves and the Gift: Kabre Sale of Kin During the Era of the Slave Trade," *Journal of African History* 37, no. 1 (1996): 31–49.

141 more than thirty million Africans: According to the *Encyclopedia Britannica's Guide to Black History*, "Approximately 18 million Africans were delivered into the Islamic trans-Saharan and Indian Ocean slave trades between 650 and 1905." Available at http://www.britannica.com/blackhistory/article-24156 (accessed June 1, 2014). In addition, Voyages: The Trans-Atlantic Slave Trade Database stated that "12.5 million embarked for the New World between 1501-1866." Available at http://www.slavevoyages.org (accessed June 1, 2014).

141 "The slaves all night": J. Iliffe, *Africans: The History of a Continent*, vol. 85 (Cambridge, UK: Cambridge University Press, 1995).

144 "When I get to heaven": Biographical information about Wantchekon came from my interview with him and from L. Wantchekon, "Dreaming Against the Grain" (unpublished).

145 **exposed to the slave trade:** N. Nunn and L. Wantchekon, "The Slave Trade and the Origins of Mistrust in Africa," *American Economic Review* 101, no. 7 (2011): 3221–52.

146 **trust was truly worthy:** They ran many additional analyses in order to rule out other possible causes of distrust, such as colonialism. As with the connection between slavery and poverty, they found that colonialism mattered but by itself could not have caused the mistrust.

146 **public meetings in Benin:** Of course, not all differences in trust or economic issues between countries can be entirely attributed to the slave trade. In Benin, for example, Wantchekon also found that trust was undermined when people joined religious sects that fostered a culture of fear.

148 **studies of the transmission of ideas:** N. Voigtländer and H.-J. Voth, "Persecution Perpetuated: The Medieval Origins of Anti-Semitic Violence in Nazi Germany," *Quarterly Journal of Economics* 127, no, 3 (2012): 1339–92.

152 **decided to test them:** A. Alesina, P. Giuliano, and N. Nunn, "On the Origins of Gender Roles: Women and the Plough," *Quarterly Journal of Economics* 128, no. 2 (2013): 469–530.

153 **like religion and language:** T. Talhelm, et al., "Large-Scale Psychological Differences Within China Explained by Rice vs. Wheat Agriculture," *Science* 344 (2012): 603–8.

154 **how many children they had:** R. Fernandez and A. Fogli, "Culture: An Empirical Investigation of Beliefs, Work, and Fertility" (working paper no. 11268, National Bureau of Economic Research, April 2005).

155 **women born in the new:** It is important to keep in mind, though, that these effects, like Nunn's trust effect and Voigtlander and Voth's legacy of hate, are probabilistic: There is no guarantee they will have an impact. There is no way to determine which child from which family will do what his grandparents did or think what his grandparents thought.

156 **"careful in dealing with people?":** G. Tabellini, "Culture and Institutions: Economic Development in the Regions of Europe," *Journal of the European Economic Association* 8, no. 4 (2010): 677–716.

156 **as well as poorer economies:** Another intuitive but actually never-before-measured observation that resulted from Nunn's analysis was the distinct geographical correlates of trust in Africa today: The closer people live to the coast, the less trusting they are. Why would this be so? Is there a relevant factor in the climate of inland Africa? Do the vagaries of fishing make you more cautious than you might be otherwise? Nunn and his colleagues looked for similar connections between coastlines and distrust in Asia and Europe, but they could find no comparable relationship.

In most parts of the world rugged, mountainous landscapes tend to be less economically successful than flatlands. It's harder to transport goods through hilly areas. The steeper a slope is, the more erosion is a problem. Watering crops is much more difficult on a slope because it's harder to control water on a hill than it is on level ground. It can simply require more effort, too, for people merely to live in this kind of landscape, let alone to harvest food or goods from it. Flatter landscapes tend to be cheaper and physically easier to farm and to move about in general. Nunn cited a report from the Food and Agriculture Organization of the United Nations that found that if a slope is greater than two degrees, then it costs more to farm it than it's worth. If it's greater than six degrees, it's not even possible to farm. Yet today in Africa, uniquely in the world, challenging landscapes have better economies. It turns out that the way trust is mapped over the African continent was also shaped by the slave trade. Before 1900, the closer you were to the coast, the more likely it was that you, your parents, and your parents' parents had been exposed in one way or another to the horror of the trade. The farther people lived from the coast, the less likely it was that they were taken as slaves. N. Nunn and D. Puga, "Ruggedness: The Blessing of Bad Geography in Africa," *Review of Economics and Statistics* 94, no. 1 (2012): 20–36.

Chapter 8: The Small Grains of History

161 **it's quite close to it**: S. Leslie, et al., "Fine Scale Genetic Structure of the British Population." Manuscript submitted for publication, 2014.

161 **"their favorite gene"**: Quotes from Peter Donnelly in this chapter are from my interviews with him.

162 **"Now we know that"**: Quotes from Stephen Leslie in this chapter are from my interviews with him.

165 **population genetics and statistics**: Most geneticists know of Fisher only as a key figure in modern genetics, while most statisticians know of him only as a key figure in statistics.

165 **more than two thousand genomes**: This was the number after ruling out people who were actually related to one another.

168 **"It was absolutely staggering"**: Quotes from Sir Walter Bodmer in this chapter are from my interviews with him.

168 **"I had naively expected"**: Quotes from Mark Robinson in this chapter are from my interviews with him.

173 **produced in Roman times**: If fourth-century Saxons invaded England today, stomped on all the cell phones, killed the engineers, and shut down the ports, how many generations would it take before stories about a small, flat object that carried people's voices in it were considered a myth?

Chapter 9: DNA + Culture

179 **"He has some very exciting news"**: Quotes from Thomas Robinson in this chapter are from my interviews with him.

180 **have had such an effect**: J. Pongratz, et al., "Coupled Climate—Carbon Simulations Indicate Minor Global Effects of Wars and Epidemics on Atmospheric CO_2 between AD 800 and 1850," *Holocene* 21, no. 5 (2011): 843–51.

181 **the ancient Mongolian empire**: T. Zerjal, et al., "The Genetic Legacy of the Mongols," *American Journal of Human Genetics* 72, no. 3 (2003): 717–21.

183 "**Most of the direct descendants**": S. Cauchi, "Descendants of Darwin Evolve into Guardians of the Wilderness," *Age*, November 29, 2009, available at http://www.theage.com.au/national/descendants-of-darwin-evolve-into -guardians-of-the-wilderness-20091128-jy10.html.

185 **eight hundred or so years:** When does the copying take place? When men and women copy their chromosomes into sex cells, the sperm and the egg. But they copy only half of their full set into each cell, twenty-three each, so that when sperm and egg get together, they make a full human complement.

188 "**It was a hellish time**": Quotes from Donald MacLaren in this chapter are from my interviews with him.

190 "**We beat on FT DNA to give us more**": Quotes from Robert McLaren in this chapter are from my interviews with him.

191 "**an old Scottish lineage**": Inheritance was not always passed directly down a single line from father to son to grandson to great-grandson. The clan leadership could pass down the male line through brothers and to their sons. But it's still the same Y chromosome. Two brothers will have the same Y as their father, and they will pass that Y on to their sons.

192 **Johns, originated this way:** M. A. Jobling, "In the Name of the Father: Surnames and Genetics," *TRENDS in Genetics* 17, no. 6 (2001): 353–57. Much of my data about British surnames, as well as the general principles, comes from interviews with Turi King and Kevin Schurer and from T. E. King and M. A. Jobling, "What's in a Name? Y Chromosomes, Surnames and the Genetic Genealogy Revolution," *TRENDS in Genetics* 25, no. 8 (2009): 351–60; and T. E. King and M. A. Jobling, "Founders, Drift, and Infidelity: The Relationship Between Y Chromosome Diversity and Patrilineal Surnames," *Molecular Biology and Evolution* 26, no. 5 (2009): 1093–1102.

193 **not the rules of language:** F. Manni, W. Heeringa, and J. Nerbonne, "To What Extent are Surnames Words? Comparing Geographic Patterns of Surname and Dialect Variation in the Netherlands," *Literary and Linguistic Computing* 21, no. 4 (2006): 507–27.

194 **Y in this population:** G. R. Bowden, et al., "Excavating Past Population Structures by Surname-Based Sampling: The Genetic Legacy of the Vikings in Northwest England," *Molecular Biology and Evolution* 25, no. 2 (2008): 301–9.

195 **non-Catholic fifth-century warlord:** My information about the Irish Y, Lord Turlough, and Irish names comes primarily from B. McEvoy and D. Bradley, "Y-Chromosomes and the Extent of Patrilineal Ancestry in Irish Surnames," *Human Genetics* 119, no. 1 (2006): 212–19; and L. T. Moore, et al., "A Y-Chromosome Signature of Hegemony in Gaelic Ireland," *American Journal of Human Genetics* 78, no. 2 (2006): 334–38.

197 **the rule does not apply:** T. E. King, et al., "Genetic Signatures of Coancestry Within Surnames," *Current Biology* 16, no. 4 (2006): 384–88.

200 "**genealogy, genetics, and . . . pure dumb luck**": Quotes from Glynis McHargue Patterson in this chapter are from my interviews with her.

200 **enough to discredit all documentation:** Note that the average nonpaternity figure will differ in different eras and within different populations, social classes, or castes.

Chapter 10: Chunks of DNA

203 **"Is this Scott Woodward?":** Quotes from Scott Woodward in this chapter are from my interview with him.

207 **just a few years ago:** Even after development of the Family Tree DNA test that looked at sixty-seven segments on the Y chromosome, many academic studies of the Y chromosome examined only seventeen segments.

207 **not at research facilities:** A. Congiu, et al., "Online Databases for mtDNA and Y Chromosome Polymorphisms in Human Populations," *Journal of Anthropological Sciences* 90 (2012): 197–212.

208 **"It's pretty boring":** Quotes from Bennett Greenspan in this chapter are from my interviews with him.

210 **"an incredibly strong woman":** Quotes from Blaine Bettinger in this chapter are from my interviews with him.

211 **"Little did I know, it's addicting":** Quotes from CeCe Moore in this chapter are from my interviews with her.

213 **networks have changed through time:** P. Ralph and G. Coop, "The Geography of Recent Genetic Ancestry Across Europe," *PLoS Biology* 11, no. 5 (2013): e1001555.

214 **second to ninth cousins:** B. M. Henn, et al., "Cryptic Distant Relatives Are Common in Both Isolated and Cosmopolitan Genetic Samples," *PLoS ONE* 7, no. 4 (2012): e34267.

215 **throughout my genealogical tree:** If I set up one of my sisters' genomes and tracked it back up the table next to mine, some of the chunks that my mother passed on to me would overlap considerably with the chunks she passed down to my sister. Yet so random is the distribution of DNA through the generations that, according to Ralph and Coop, we may share no DNA with a number of our fourth cousins. It's possible, though less likely, that we won't share any DNA with some third cousins. In fact, technically it's possible that we have no DNA from one of our grandparents, but, according to Ralph, the chance of that happening is roughly one in ten trillion. See also R. Khan, "Which Grandparent Are You Most Related To?" *Slate*, October 18, 2013, available at http://www .slate.com/articles/health_and_science/human_genome/2013/10/analyze _your_child_s_dna_which_grandparents_are_most_genetically_related.html.

217 **"It took us a while":** Quotes from Peter Ralph in this chapter are from my interviews with him.

217 **1,000 and 1,500 years ago:** Ralph's exercise made me wonder what would happen if I traced my genome back through, say, fifty thousand years. In a nutshell, it would be very disintegrated and widely spread. But what if we started from that point fifty thousand years ago and then asked: What are the odds that all those minuscule bits that are spread all over humanity will come together to form a single genome in fifty thousand years? Calculating that is pretty much impossible. Yet somehow it happened.

217 **This has implications for the genetic databases:** Even though sharing a common segment with someone doesn't necessarily mean you have inherited that block from a relatively recent common ancestor, it is often the case that

when genetic genealogy companies help customers with shared DNA hook their genealogical trees together, they do find a distant shared cousin from, say, six generations ago. What is the likelihood in this case that the shared block comes not from a shared fourth-great-grandparent but in fact from someone much further back along a different route through the genealogical tree? According to Ralph, the chance of two people sharing a block from a sixth-generation ancestor is close to one, so the chance of the shared block coming from an identified ancestor is pretty high. Yet it is still possible that those two people also share a block that doesn't come from that ancestor. "If the known ancestor was Charlemagne," Ralph explained, "the chance that you both inherited blocks from him is pretty small, so the block is probably not from him, especially given the huge number of unknown relatives you share from the time of Charlemagne."

219 **contributed nothing to you genetically**: The effect was first pointed out to Patterson by Oxford professor Jotun Hein in 2004. See D. L. Rohde, S. Olson, and J. T. Chang, "Modelling the Recent Common Ancestry of All Living Humans," *Nature* 431, no. 7008 (2004): 562–66.

223 **"Northern Italy had the tradition"**: Quotes from Guido Tabellini in this chapter are from my interview with him.

Chapter 11: The Politics of DNA

226 **"an intimate and loving relationship"**: G. Wood, "The Sally Hemings Case," Barbara Chase-Riboud, reply by Gordon S Wood, *New York Review of Books*, June 12, 1997, available at http://www.nybooks.com/articles/archives/1997/jun/12/the-sally-hemings-case/. Note that in a May 2013 *New York Review of Books* article, Wood praised Gordon-Reed's acute and correct analysis.

227 **match the Hemings Y**: E. A. Foster, et al., "Jefferson Fathered Slave's Last Child," *Nature* 396, no. 6707 (1998): 27–28.

227 **deny the Jefferson/Hemings link**: E. S. Lander and J. J. Ellis, "Founding Father," *Nature* 396, no. 6707 (1998): 13–14.

227 **"suggests the strong likelihood"**: D. P. Jordan, Statement on the TJMF Research Committee Report on Thomas Jefferson and Sally Hemings, available at http://www.monticello.org/site/plantation-and-slavery/report-research-committee-thomas-jefferson-and-sally-hemings.

228 **"veins was Thomas Jefferson's"**: M. Hendricks, "A Daughter's Dedication," *Johns Hopkins Magazine*, September 1999, available at http://pages.jh.edu/~jhumag/0999web/roots.html.

229 **stories and the DNA evidence**: S. R. Williams, "Genetic Genealogy: The Woodson Family's Experience," *Culture, Medicine and Psychiatry* 29, no. 2 (2005): 225–52; and personal communciation.

229 **"They matched so well"**: Unless otherwise cited, quotes from Sloan Williams in this chapter are from my interviews with her.

229 **"They were extremely suspicious"**: S. R. Williams, "Genetic Genealogy: The Woodson Family's Experience," *Culture, Medicine and Psychiatry* 29, no. 2 (2005) 226.

231 **black—were as well:** CeCe Moore of YourGeneticGenealogist.com currently runs an autosomal DNA project to gather DNA from descendants of Madison Hemings (as well as descendants of Eston and the other Hemings children) in order to try to match it against descendants of the known Jefferson lineage. She also accepts DNA from other nonlineal relatives of the Jefferson and Hemings families.

233 **"results are unexpected or undesired":** D. A. Bolnick, et al., "The Science and Business of Genetic Ancestry Testing," *Science* 318, no. 5849 (2007): 399.

233 **"general and the scientific communities":** C. D. Royal, et al., "Inferring Genetic Ancestry: Opportunities, Challenges, and Implications," *American Journal of Human Genetics* 86, no. 5 (2010): 661–73.

233 **"more problems than it solves":** C. Elliott and P. Brodwin, "Identity and Genetic Ancestry Tracing," *British Medical Journal* 325, no. 7378 (2002): 1469.

233 **few scientists would disagree:** J. Marks, "Contemporary Bio-Anthropology," *Anthropology Today* 18, no. 4 (2002)): 3, 5, 7; and J. Marks, "'We're Going to Tell These People Who They Really Are': Science and Relatedness," in *Relative Values: Reconfiguring Kinship Studies*, ed. S. Franklin and S. McKinnon (Durham and London: Duke University Press, 2001), pp. 355–83.

234 **"Africa and Africans as primordial":** K. TallBear, "Narratives of Race and Indigeneity in the Genographic Project," *The Journal of Law, Medicine & Ethics* 35, no. 3, 412–424.

236 **"political agenda of science haters":** C. Tuniz, R. Gillespie, and C. Jones, *The Bone Readers* (Crows Nest: Allen & Unwin, 2009), 195

236 **differently shaped skulls:** L. Betti, et al., "The Relative Role of Drift and Selection in Shaping the Human Skull," *American Journal of Physical Anthropology* 141, no. 1 (2010): 76–82.

236 **no such thing as biological race:** Lewontin is not the only researcher who made this case, yet while others made it before him, he is most often popularly associated with this argument today.

237 **"Human races and populations are remarkably similar":** R. C. Lewontin, "The Apportionment of Human Diversity," *Evolutionary Biology* 6 (1972): 381–98. For a recent essay about Lewontin's views on race and ancestry see R. Lewontin, "Confusions About Human Races," *Is Race "Real"?* June 7, 2006, available at http://raceandgenomics.ssrc.org/Lewontin/

237 **"two random individuals":** D. Witherspoon, "Genetic Similarities Within and Between Human Populations," *Genetics*, May 2007; 176 (1): 351.

237 **the picture changes:** This was first pointed out in A. W. Edwards, "Human Genetic Diversity: Lewontin's Fallacy," *BioEssays* 25, no. 8 (2003): 798–801. Also see N. Risch, et al., "Categorization of Humans in Biomedical Research: Genes, Race and Disease," *Genome Biology* 3, no. 7 (2002): 1–12.

237 **than from a distant one:** D. Witherspoon, "Genetic Similarities Within and Between Human Populations," *Genetics*, May 2007; 176 (1): 351.

239 **genetics as a solution to disease:** L. Braun, "Reifying Human Difference: The Debate on Genetics, Race, and Health," *International Journal of Health Services* 36, no. 3 (2006): 557–73.

239 **medical utility of "race":** A. M. Leroi, "A Family Tree in Every Gene," *Journal of Genetics* 84, no. 1 (2005): 3–6.

239 **"clear benefits for public health"**: J. Stevens, "Eve Is from Adam's Rib, the Earth Is Flat, and Races Come from Genes," in *Is Race "Real"?* June 7, 2006, available at http://raceandgenomics.ssrc.org/Stevens/.

240 **"data-rich scientists"**: Quotes from Eran Elhaik in this chapter are from my interviews with him.

240 **"You cannot do it for every population"**: A different study found that Indian castes seemed to differ in the relative proportion of northern and southern Indian DNA in their genome. D. Reich, et al., "Reconstructing Indian Population History," *Nature* 461, no. 7263 (2009): 489–94.

241 **Ancestry does not work that way**: Other analyses have shown that the racial/ethnic self-identification of people who were sorted into clusters based on genomic markers agreed with the clustering. See, for example, N. Risch, et al., "Categorization of Humans in Biomedical Research: Genes, Race and Disease," *Genome Biology* 3, no. 7 (2002): 1–12.

242 **a survey conducted by Wendy Roth**: W. D. Roth and B. Ivemark, "'Not Everybody Knows That I'm Actually Black': The Effects of DNA Ancestry Testing on Racial and Ethnic Boundaries," presented at the American Sociological Association Annual Meeting Atlanta, August 14–17, 2010.

242 **"general level of ignorance"**: Unless otherwise cited, quotes from Wendy Roth in this chapter are from my interviews with her.

242 **to include the new information**: Their identification depended on a number of factors, including their level of education, how distant they felt from the group in question, whether they looked physically as if they belonged to one group and not another, and how their family identified itself.

244 **"integrating the contributing factors"**: Quotes from Jennifer Wagner in this chapter are from my interviews with her.

244 **"the study of peas or fruit flies"**: N. G. Jablonski, M. Shriver, and H. Gates, "Using Genetics and Genealogy to Teach Evolution and Human Diversity," National Evolutionary Synthesis Center, Catalysis Meeting, available at https://www.nescent.org/science/awards_summary.php?id=321.

Chapter 12: The History of the World

246 **intentional engraving were discovered**: P.-J. Texier, et al., "A Howiesons Poort Tradition of Engraving Ostrich Eggshell Containers Dated to 60,000 Years Ago at Diepkloof Rock Shelter, South Africa," *Proceedings of the National Academy of Sciences* 107, no. 14 (2010): 6180–85.

247 **more than 100,000 years ago**: T. F. Strasser, et al., "Dating Palaeolithic Sites in Southwestern Crete, Greece," *Journal of Quaternary Science* 26, no. 5 (2011): 553–60.

247 **ocher processing**: C. S. Henshilwood, et al., "A 100,000-Year-Old Ochre-Processing Workshop at Blombos Cave, South Africa," *Science* 334, no. 6053 (2011): 219–22.

247 **their stone tools using heat**: V. Mourre, P. Villa, and C. S. Henshilwood, "Early Use of Pressure Flaking on Lithic Artifacts at Blombos Cave, South Africa," *Science* 330, no. 6004 (2010): 659–62.

247 **perhaps not much more than one thousand:** The number of individuals in
     an ancestral population is usually considerably greater than the number of
     individuals who reproduce, known as the effective population size. The esti-
     mate of 1,000-2,500 individuals is the effective population size. See B. Henn,
     et al. "The Great Human Expansion," *Proceedings of the National Academy of
     Sciences* 109, 44 (2012): 17758–17764.

247 **"that happened a long time ago":** Quotes from Marcus Feldman in this
     chapter are from my interview with him.

249 **skating on ice:** F. Formenti and A. E. Minetti, "The First Humans Travelling
     on Ice: An Energy-Saving Strategy?" *Biological Journal of the Linnean Soci-
     ety* 93, no. 1 (2008): 1–7.

249 **Pacific coastline and then eastward:** J. A. Raff and D. A. Bolnick, "Ge-
     netic Roots of the First Americans," *Nature* 506, no. 7487 (2014): 162–63.

250 **Eurasia that was ancestral to both:** N. Patterson, et al., "Ancient Admix-
     ture in Human History," *Genetics* 192, no. 3 (2012): 1065–93.

250 **population in western Eurasia:** M. Raghavan, et al., "Upper Palaeolithic
     Siberian Genome Reveals Dual Ancestry of Native Americans," *Nature* 505,
     no. 7481 (2014): 87–91.

250 **others from North America:** M. Rasmussen, et al., "The Genome of a Late
     Pleistocene Human from a Clovis Burial Site in Western Montana," *Nature*
     506, no. 7487 (2014): 225–29.

250 **wild canine in with them:** I. Pugach, et al., "Genome-wide Data Substanti-
     ate Holocene Gene Flow from India to Australia," *Proceedings of the National
     Academy of Sciences* 110, no. 5 (2013): 1803–8.

251 **blended with other groups:** C. M. Schlebusch, et al., "Genomic Variation in
     Seven Khoe-San Groups Reveals Adaptation and Complex African History,"
     Science 338, no. 6105 (2012): 374–79.

251 **modern human exodus from Africa:** H. Reyes-Centeno, et al., "Genomic
     and Cranial Phenotype Data Support Multiple Modern Human Dispersals
     from Africa and a Southern Route into Asia," *Proceedings of the National
     Academy of Sciences* 111, no. 20, (2014): 7248–53.

251 **years ago support this idea:** J. Rose, et al., "The Nubian Complex of Dhofar,
     Oman: An African Middle Stone Age Industry in Southern Arabia," *PLoS
     ONE* 6, no. 11 (2011): e28239.

252 **"Neanderthals and *Homo sapiens* are like lions and tigers":** Quote
     from Colin Groves in this chapter is from my interview with him. Razib Khan
     suggested to me that the difference was more like polar bears versus brown
     bears.

252 **including David Reich at Harvard:** R. E. Green, et al., "A Draft Sequence
     of the Neandertal Genome," *Science* 328, no. 5979 (2010): 710–22.

253 **the beginning of metallurgy:** G. Brandt, et al., "Ancient DNA Reveals Key
     Stages in the Formation of Central European Mitochondrial Genetic Diver-
     sity," *Science* 342, no. 6155 (2013): 257–61.

253 **farmers replaced the hunter-gatherers:** P. Skoglund, et al., "Genomic Di-
     versity and Admixture Differs for Stone-Age Scandinavian Foragers and
     Farmers," *Science* 344 no. 6185 (2014): 747–750.

254 **such as addiction to cigarettes:** S. Sankararaman, et al., "The Genomic Landscape of Neanderthal Ancestry in Present-Day Humans," *Nature* 507, no. 7492 (2014): 354–57.

254 **lipid catabolism:** E. Khrameeva, "Neanderthal Ancestry Drives Evolution of Lipid Catabolism in Contemporary Europeans," *Nature Communications* 5, available at DOI: 10.1038/ncomms4584.

254 **the genomes of modern people:** J. Krause, et al., "The Complete Mitochondrial DNA Genome of an Unknown Hominin from Southern Siberia," *Nature* 464, no. 7290 (2010): 894–97.

255 **different, as yet unknown, species:** M. F. Hammer, et al., "Genetic Evidence for Archaic Admixture in Africa," *Proceedings of the National Academy of Sciences* 108, no. 37 (2011): 15123–28.

258 **several times over in different groups:** S. A. Tishkoff, et al., "Convergent Adaptation of Human Lactase Persistence in Africa and Europe," *Nature Genetics* 39, no. 1 (2007): 31–40.

259 **the better we could process starch:** G. Perry, "Diet and the Evolution of Human Amylase Gene Copy Number Variation," *Nature Genetics* 39 (2007): 1256–1260.

260 **a series of apocalyptic Mexican pandemics:** R. Acuna-Soto, et al., "Megadrought and Megadeath in 16th Century Mexico," *Revista Biomédica* 13 (2002): 289–292.

261 **New Guinea/Australia bacteria:** Y. Moodley, et al., "The Peopling of the Pacific from a Bacterial Perspective," *Science* 323, no. 5913 (2009): 527–30.

262 **based on mouse movements alone:** J. B. Searle, et al., "Of Mice and (Viking?) Men: Phylogeography of British and Irish House Mice," *Proceedings of the Royal Society B: Biological Sciences* 276, no. 1655 (2009): 201–7.

262 **did not leave a lasting imprint:** E. P. Jones, et al., "Fellow Travellers: A Concordance of Colonization Patterns Between Mice and Men in the North Atlantic Region," *BMC Evolutionary Biology* 12, no. 1 (2012): 35.

Chapter 13: The Past Is Written on Your Face: DNA, Traits, and What We Make of Them

267 **"One of the most fascinating mysteries in Tennessee lore":** W. Winkler, *Walking Toward the Sunset: The Melungeons of Appalachia* (Macon, GA: Mercer University Press, 2005), ix.

268 **"explain what a Melungeon is":** Wagne Winkler's quote comes from my interviews with him. Unless otherwise cited as W. Winkler, *Walking Toward the Sunset: The Melungeons of Appalachia* (Macon, GA: Mercer University Press, 2005); and W. Winkler, "Melungeons Yesterday and Today: Thirty Years Later," Melungeon Heritage Association, 2005, available at http://melungeon.ning.com/forum/topics/2005-winkler-article-on-jean-patterson-bible-s-study-of (accessed April 17, 2014), then other Winkler quotes come from my interviews with him.

270 **"A Melungeon isn't":** J. Bible, *Melungeons Yesterday and Today* (Signal Mountain, TN: Mountain Press 5th ed., 1975), 13.

270 "After the breaking out of the war": W. Winkler, *Walking Toward the Sunset: The Melungeons of Appalachia* (Macon, GA: Mercer University Press, 2005, 261), 271.

272 Kennedy, who wrote *The Melungeons*: N. B. Kennedy and R. V. Kennedy, *The Melungeons: The Resurrection of a Proud People: An Untold Story of Ethnic Cleansing in America* (Macon, GA: Mercer University Press, 1997). As outlandish as the idea seems that some Melungeons may have descended from settlers who arrived in the United States before Jamestown, or even who came from Roanoke, it's not that different from another legend-turned-fact further up the East Coast. The stories that Vikings sailed to North America long before the Spanish or the British were considered a fantasy until the 1960s, when the remains of a Norse settlement were dug up in Newfoundland.

272 "So smooth of tongue": W. Winkler, *Walking Toward the Sunset: The Melungeons of Appalachia* (Macon, GA: Mercer University Press, 2005), 7.

275 I want to document as best I can: W. Winkler, *Walking Toward the Sunset: The Melungeons of Appalachia* (Macon, GA: Mercer University Press, 2005), xii.

276 "I saw the still-living tentacles": N. B. Kennedy and R. V. Kennedy, *The Melungeons: The Resurrection of a Proud People: An Untold Story of Ethnic Cleansing in America* (Macon, GA: Mercer University Press, 1997), 7.

276 late nineteenth and early twentieth centuries: This study's method of selecting subjects was similar to the Viking surname and the Y chromosome study. By choosing names that were independently associated with Viking history and then examining subjects who bore those names for a Viking Y, the researchers in that study maximized the chance that they would find a historical trace of Viking men. In the Melungeon study, by using one of the only independent records of Melungeons that exist, the researchers were able to narrow their focus much more effectively.

278 people have shovel-shaped incisors: Finnish people have shovel-shaped incisors too. It's thought that as the ancient Asian travelers moved into Beringia, they left some genes in the Finnish Sami population.

278 "For some reason, people": Quotes from Richard Scott in this chapter are from my interview with him.

278 single letter within a single gene: Variation in the trait is affected by a single letter of DNA; for example, if a particular spot in the gene is filled by a C, then the carrier has wet earwax. If instead the spot is filled by a T, the carrier will have dry earwax. Customers of 23andMe can find out which earwax gene they have. They can also find out if they have the gene that controls the flush reaction to alcohol and the gene that controls the ability to taste bitter tastes, among others.

278 way from the full picture: R. Kimura, et al., "A Common Variation in EDAR Is a Genetic Determinant of Shovel-Shaped Incisors," *American Journal of Human Genetics* 85, no. 4 (2009): 528–35.

280 more than ten thousand Europeans: F. Liu, et al., "A Genome-Wide Association Study Identifies Five Loci Influencing Facial Morphology in Europeans," *PLoS Genetics* 8, no. 9 (2012): e1002932.

280 **They found five genes:** PAX3, TP63, and one other gene had already been implicated in other studies, but the association of the remaining two with the face was completely new.

281 **DNA on the structure of the face:** C. Attanasio, et al., "Fine Tuning of Craniofacial Morphology by Distant-Acting Enhancers," *Science* 342, no. 6157 (2013): 1241006.

281 **reconstruction of faces from ancient remains:** J. Draus-Barini, et al., "Bona Fide Colour: DNA Prediction of Human Eye and Hair Colour from Ancient and Contemporary Skeletal Remains," *Investigative Genetics* 4, no. 1 (2013): 3.

281 **forensic police profiling:** P. Claes, et al., "Modeling 3D Facial Shape from DNA," *PLoS Genetics* 10, no. 3 (2014): e1004224.

282 **"The fact that identical twins":** Quotes from Walter Bodmer in this chapter are from my interview with him.

283 **the accused is of a different race:** C. A. Meissner and J. C. Brigham, "Thirty Years of Investigating the Own-Race Bias in Memory for Faces: A Meta-Analytic Review," *Psychology, Public Policy, and Law* 7, no. 1 (2001): 3.

284 **African, and East Asian ancestry:** Y. C. Klimentidis and M. D. Shriver, "Estimating Genetic Ancestry Proportions from Faces," *PloS ONE* 4, no. 2 (2009): e4460.

285 **results from just such a mutation:** R. M. Harding, et al., "Evidence for Variable Selective Pressures at MC1R," *American Journal of Human Genetics* 66, no. 4 (2000): 1351–61; and P. R. John, et al., "DNA Polymorphism and Selection at the Melanocortin-1 Receptor Gene in Normally Pigmented Southern African Individuals," *Annals of the New York Academy of Sciences* 994, no. 1 (2003): 299–306.

Chapter 14: The Past May Not Make You Feel Better:
DNA, History, and Health

286–87 **Cindy Carroll was in her midforties . . . to him it felt like hours:** L. Priest, "'I Know How I Am Going to Die,'" *Globe and Mail*, October 13, 2007, available at http://www.theglobeandmail.com/life/i-know-how-i-am-going-to-die /article1084238/?page=all (accessed April 24, 2014).

288 **called *el mal* or "the bad":** M. S. Okun and N. Thommi, "Americo Negrette (1924 to 2003): Diagnosing Huntington Disease in Venezuela," *Neurology* 63, no. 2 (2004): 340–43.

288 **"strange movements, like dancing":** R. Weiser, "Huntington's Disease: A View of Maracaibo Lake" (lecture, World Congress on Huntington's Disease, Rio de Janeiro, September 16, 2013), available at http://vimeo.com/75658670.

290 **"We just learned the alphabet":** Unless otherwise cited, quotes from Jeff Carroll in this chapter are from my interview with him.

295 **"seem to be very close":** Unless otherwise cited, quotes from Feldman in this chapter are from my interview with him.

297 **up to 10 percent of all humans:** A. Bittles and M. Black, "Consanguinity, Human Evolution and Complex Diseases," *Proceedings of the National Academy of Sciences* 107, no. 1 (2010): 1779–1786.

299 speculated that they must have been: G. McDowell, et al., "The Presence
     of Two Different Infantile Tay-Sachs Disease Mutations in a Cajun Popula-
     tion," *American Journal of Human Genetics* 51, no. 5 (1992): 1071–77.

300 three-thousand-year-old culture: When they left their homeland, they also
     committed to the sect's strict practices. For example, when menstruating or
     after childbirth, they—along with all women—are considered unclean. They
     are isolated during this time and not allowed to touch anyone, even their own
     children, for the first seven days of a period and for forty days after the birth of
     a son and eighty days after the birth of a daughter.

300 "women outside our community": T. Heneghan, "Samaritans Use Modern
     Means to Keep Ancient Faith," *Reuters*, June 2, available at http://mobile.re
     uters.com/article/idUSTRE55201720090603?irpc=932.

302 "This is my wife and she is my niece": Unless otherwise cited, quotes from
     Alan Bittles in this chapter are from my interview with him.

303 the health of individuals today: B. M. Henn, et al., "Hunter-Gatherer Ge-
     nomic Diversity Suggests a Southern African Origin for Modern Humans," *Pro-
     ceedings of the National Academy of Sciences* 108, no. 13 (2011): 5154–62.

304 Why? They don't know: T. Manolio, et al., "Finding the Missing Heritability
     of Complex Diseases," *Nature* 461, 7265 (2009): 747–753.

305 did not experience such conditioning: B. G. Dias and K. J. Ressler, "Pa-
     rental Olfactory Experience Influences Behavior and Neural Structure in Sub-
     sequent Generations," *Nature Neuroscience* 17, no. 1 (2014): 89–96.

305 from Greenland, was published: M. Rasmussen, et al., "Ancient Human
     Genome Sequence of an Extinct Palaeo-Eskimo," *Nature* 463, no. 7282 (2010):
     757–62.

305 passed down experiences and predispositions: D. Gokhman, et al., "Re-
     constructing the DNA Methylation Maps of the Neandertal and the Deniso-
     van," *Science* 344, no. 6183 (2014): 523–27.

307 "Huntington's disease has been": Unless otherwise cited, quotes from Rob-
     ert Green in this chapter are from my interviews with him.

308 "a disease as frightening and untreatable": R. C. Green, et al., "Disclo-
     sure of APOE Genotype for Risk of Alzheimer's Disease," *New England Jour-
     nal of Medicine* 361, no. 3 (2009): 245–54.

311 "Historical research has shown that the idea": American Anthropological
     Association Statement on "Race," May 17, 1998, available at http://www
     .aaanet.org/stmts/racepp.htm.

Epilogue

314 decision making in India: K. R. Hoff, M. Kshetramade, and E. Fehr, "Caste
     and Punishment: The Legacy of Caste Culture in Norm Enforcement," IZA
     Discussion Paper no. 4343, August 2009.

315 "Men still didn't like women leaders": Unless otherwise cited, quotes from
     Karla Hoff in this chapter are from my interview with her.

316 "information wants to be free": S. Pinker, "My Genome, My Self," *New York
     Times Magazine*, January 11, 2009.

317 **"No harm can come"**: Unless otherwise cited, quotes from Steven Pinker in this chapter are from my interview with him.

317 **a professor at Duke University**: M. Angrist, *Here Is a Human Being: At the Dawn of Personal Genomics* (New York: Harper Collins, 2010).

317 **"One thing I notice is alleles"**: Unless otherwise cited, quotes from Misha Angrist in this chapter are from my interview with him.

317 **"It's like learning American history"**: Quotes from Esther Dyson in this chapter are from my interview with her.

318 **they found their families too**: M. Gymrek, et al., "Identifying Personal Genomes by Surname Inference," *Science* 339, no. 6117 (2013): 321–24.

319 **passed through a language barrier**: R. M. Ross, S. J. Greenhill, and Q. D. Atkinson, "Population Structure and Cultural Geography of a Folktale in Europe," *Proceedings of the Royal Society B: Biological Sciences* 280, no. 1756 (2013): 2012.3065.

319 **whereas genes are creamy**: R. Khan, "Why Culture Is Chunky and Genes Are Creamy," *Gene Expression*, February 6, 2013, available at http://blogs.discover magazine.com/gnxp/2013/02/why-culture-is-chunky-and-genes-are-creamy/? utm_source=feedburner&utm_medium=feed&utm_campaign=Feed%3A+Ge neExpressionBlog+%28Gene+Expression%29#.U5NSwJRdUsy.

320 **of course, Shakespeare's sonnets**: N. Goldman, et al., "Towards Practical, High-Capacity, Low-Maintenance Information Storage in Synthesized DNA," *Nature* 494, no. 7435 (2013).

INDEX